THE LIBRARY
ST. MARY'S COLLEGE OF MARYLAND
ST. MARY'S CITY, MARYLAND 20686

About Island Press

Island Press is the only nonprofit organization in the United States whose principal purpose is the publication of books on environmental issues and natural resource management. We provide solutions-oriented information to professionals, public officials, business and community leaders, and concerned citizens who are shaping responses to environmental problems.

In 2000, Island Press celebrates its sixteenth anniversary as the leading provider of timely and practical books that take a multidisciplinary approach to critical environmental concerns. Our growing list of titles reflects our commitment to bringing the best of an expanding body of literature to the environmental community throughout North America and the world.

Support for Island Press is provided by The Jenifer Altman Foundation, The Bullitt Foundation, The Mary Flagler Cary Charitable Trust, The Nathan Cummings Foundation, The Geraldine R. Dodge Foundation, The Charles Engelhard Foundation, The Ford Foundation, The Vira I. Heinz Endowment, The William and Flora Hewlett Foundation, The W. Alton Jones Foundation, The John D. and Catherine T. MacArthur Foundation, The Andrew W. Mellon Foundation, The Charles Stewart Mott Foundation, The Curtis and Edith Munson Foundation, The National Fish and Wildlife Foundation, The National Science Foundation, The New-Land Foundation, The David and Lucile Packard Foundation, The Pew Charitable Trusts, The Rockefeller Brothers Fund, Rockefeller Financial Services, The Surdna Foundation, The Winslow Foundation, and individual donors.

Nature's Fading Chorus

Nature's Fading Chorus
Classic and Contemporary Writings on Amphibians

Edited by Gordon L. Miller

ISLAND PRESS
Washington, D.C. • Covelo, California

Copyright © 2000 by Gordon L. Miller

All rights reserved under International and Pan-American Copyright Conventions. No part of this book may be reproduced in any form or by any means without permission in writing from the publisher: Island Press, 1718 Connecticut Avenue, N.W., Suite 300, Washington, DC 20009.

ISLAND PRESS is a trademark of The Center for Resource Economics.

A list of permissions and original sources appears on pages 223–225.

Library of Congress Cataloging-in-Publication Data
Nature's fading chorus : classic and contemporary writings on amphibians / edited by Gordon L. Miller.
    p.   cm.
  Includes bibliographical references and index.
    ISBN 1–55963–793–5 (cloth : acid-free paper)—ISBN 1–55963–794–3 (paper : acid-free paper)
    1. Amphibians. I. Miller, Gordon L., 1954–
    QL667 .N38   2000
    597.8—dc21
                                               00–008242
                                                    CIP

Printed on recycled, acid-free paper

Manufactured in the United States of America
10 9 8 7 6 5 4 3 2 1

Contents

Prologue *by Robert Michael Pyle* *xi*
Introduction *1*

Part I: Interpreting the Cosmos *11*
Early Naturalists

ARISTOTLE *14*
From *Historia Animalium (Inquiry Concerning Animals)* (fourth century B.C.E.)

PLINY THE ELDER *16*
From *Natural History* (first century C.E.)

ANONYMOUS *18*
From the *Physiologus* (ca. second century C.E.)

ALBERT THE GREAT *19*
From *De Animalibus (Man and the Beasts)* (ca. 1250)

EDWARD TOPSELL *22*
From *The History of Four-Footed Beasts and Serpents* (1608)

THOMAS BROWNE *26*
From *Pseudodoxia Epidemica* (1646)

JAN SWAMMERDAM *31*
From *The Book of Nature; or, The History of Insects* (1669)

JOHN RAY *33*
From *The Wisdom of God Manifested in the Works of the Creation* (1691)

Part II: Reclaiming Paradise *39*
Pioneering Nature Writers

GILBERT WHITE *41*
From *The Natural History of Selborne* (1789)

WILLIAM BARTRAM *45*
From *Travels Through North and South Carolina, Georgia, East and West Florida* (1791)

HENRY DAVID THOREAU *50*
From the *Journal* (1857–1860)

JOHN BURROUGHS *55*
"The Tree-Toad" (1904)

W. H. HUDSON *59*
"The Toad as Traveller" (1919)

Part III: Telling Naturalistic Tales *65*
Scientific Essayists

CHARLES DARWIN *68*
From *The Voyage of the Beagle* (1845) and *The Origin of Species* (1859)

THOMAS HENRY HUXLEY *72*
"On the Hypothesis That Animals Are Automata, and Its History" (1874)

J. ARTHUR THOMSON *76*
"The Tale of Tadpoles" (1911)

JULIAN HUXLEY *81*
"The Frog and Biology" (1927)

LOREN EISELEY *88*
"The Dance of the Frogs" (1978)

STEPHEN JAY GOULD *95*
"Here Goes Nothing" (1991)

DAVID SCOTT   *105*
"A Breeding Congress" (1998)

Part IV: Remembering the Earth   *111*
Twentieth-Century Nature Writers

DALLAS LORE SHARP   *113*
From *The Face of the Fields* (1911) and *Sanctuary! Sanctuary!* (1926)

DONALD CULROSS PEATTIE   *124*
From *An Almanac for Moderns* (1935)

GEORGE ORWELL   *128*
"Thoughts on the Common Toad" (1946)

JOSEPH WOOD KRUTCH   *131*
"The Day of the Peepers" (1949)

EDWIN WAY TEALE   *136*
"Audubon's Salamanders" (1965)

ANNIE DILLARD   *142*
From *Pilgrim at Tinker Creek* (1974)

ROBERT MICHAEL PYLE   *146*
From *Wintergreen* (1986)

ANN HAYMOND ZWINGER   *153*
From *The Mysterious Lands* (1989) and *Downcanyon* (1995)

TERRY TEMPEST WILLIAMS   *157*
From *Desert Quartet* (1995)

Part V: Reading the Signs of the Times   *161*
Declines, Deformities, and Biodiversity

KATHRYN PHILLIPS   *164*
From *Tracking the Vanishing Frogs* (1994)

LAURIE J. VITT, JANALEE P. CALDWELL, HENRY M. WILBUR, AND
DAVID C. SMITH   *175*
"Amphibians as Harbingers of Decay" (1990)

ANDREW R. BLAUSTEIN   *176*
"Amphibians in a Bad Light" (1994)

TIMOTHY R. HALLIDAY AND W. RONALD HEYER   *181*
"The Case of the Vanishing Frogs" (1997)

STEPHEN LEAHY   *189*
"The Sound of Silence" (1998)

VIRGINIA MORELL   *194*
"Are Pathogens Felling Frogs?" (1999)

JOCELYN KAISER   *200*
"A Trematode Parasite Causes Some Frog Deformities" (1999)

GORDON L. MILLER   *203*
"Dimensions of Deformity"

Epilogue *by Ann Haymond Zwinger*   *215*
Notes   *219*
Sources   *223*
Acknowledgments   *227*
About the Authors   *229*
Index   *237*

Prologue
Reflections in a Golden Eye

> "Who knows whether these broken heavens
> Could exist tonight separate from trills and toad ringings?"
> —Pattiann Rogers, *The Power of Toads*

Last summer, my wife, Thea, and I were looking for border butterflies along the shores of Clear Lake, in the foothills of the North Cascade in Washington State, just below British Columbia. I was following an unseasonably late Sara orangetip to make sure of its identity. Thea had just spotted the first dun skipper ever recorded in Whatcom County.

I heard her call. "Oh!" she cried. "Come quick!" I ran back to find her staring into a ditch, incredulity like the hot sun on her face. I looked, too, and saw what she saw—thousands of big bugs crawling out of the dry ditch and heading in all directions.

At least, I took them for insects at first. "Geez! Crickets!" I said. The whole ditch and roadside were alive with the creatures. But then we both looked at them more closely, and when our shock gave way to clear eyes, we saw what they really were: toads!

Toadlets, actually—each one an inch in length or half that, not long from the tadpole. In fact, and now we remembered, we had recently seen the nearby lakeshore brimming with polliwogs, but at the time, we hadn't known they were toads. Now, to stand there and behold tiny, scurrying, squirming hoppytoads by the many, many hundreds—well, we were thrilled.

Time was, one could go for a walk in the Cascades and be reasonably sure of encountering a few northern toads. *Bufo boreas* was a regular, even common resident of most of the moist and montane Maritime Northwest. But not any more. Now, the rest of Washington State seems much like the Willapa Hills, where my wife and I live, which seem always to have been essentially absent of toads. With the exception of a few little ones hanging around the lights of the comfort station at Keller Ferry on the Columbia River a couple of years ago, this mass emergence was the first sign of them I'd seen in years. And when I'd attended a meeting of the Northwestern Vertebrate Society in Astoria, Oregon, not long back, one of the presented papers had carried this shocking title: "One More Nail in the Coffin of the Northern Toad."

Toads. I had always been fond of them. My first acquaintance with them had probably been through the Thornton Burgess Bedtime Stories, which doubled for me as storybooks and natural history texts all through my early youth. Burgess's animals may have worn waistcoats and top hats, but they also did the things those animals really do. Old Mr. Toad, for example, shoots out his tongue for fat, green flies; goes hop, hop, hipperty-hop down to the Smiling Pool to take part in the spring chorus; and shows Peter Rabbit how he can disappear by digging his way into the sand for the winter. Harrison Cady's broad-mouthed and bumpy portrait still defines the species for me more than any field guide mug shot. The adventures of Mr. Toad, along with those of Paddy the Beaver, Billy Otter, Mr. Blacksnake, and Chatterer the Red Squirrel, constituted my sole experience with toads in the wild for some years, as they were missing from our dry, new gardens on the unmade edges of the Denver suburbs.

Then there came that other Mr. Toad, the one with the motorcar that went *Poop! Poop!* Admittedly, I saw the Disney version of *The Wind in the Willows* before I ever read Kenneth Grahame's enchanting book. Perhaps Toad's misadventures on the open road do not exactly mirror natural history. But I read of Mole and Ratty finding the baby otter safe on the reedy island in the river, in the protective care of the Piper at the Gates of Dawn, it surely lured me out to my own local watercourse to seek such things for myself. In the intermittent ditch that was my Valhalla, amphibians proved as elusive as otters or the land snails I so keenly coveted. Yet even if I never found a toad on the High Line Canal, one would nevertheless play a role in telling the story of that prairie "crick" many years later. In my book *The Thunder Tree,* I brought a *Bufo* to the High Line in the form of an inadvertent passenger and observer. There, a semifictional narrative describes the entire length of the canal through the eyes of this vagrant toad, who hopped onto a mat of moss in the high country and then washed down the Platte River and into the canal in the spring runoff.

The idea for the toad's journey arose from my first real encounter with *Bufo boreas*. In the summer of 1964, most of my friends were working at dreadful jobs in Denver for $0.75 per hour. To escape into the mountains, several of us took an even more dreadful job that my older brother Tom had found, clearing trees for power line right-of-way at an elevation of 10,000 feet in the Rocky Mountains, way up above Georgetown, for $1.25 per hour. Because we had to show up for the backbreaking work at six in the morning, we just camped nearby, on the edge of an alpine bog. One day after the grinding work, I was prowling the willow-fragrant sphagnum moss when I spotted a small western toad. Two weeks later (when we'd had all of that job we could stand), we left for home, and I took the toad with me. I named it Todd, after one of my workmates, kept it for some weeks in a mossy terrarium, and then released it in its home habitat before the frost settled in. Todd served as Herpetology 101 for me and initiated a lifelong love affair with the real thing.

When I went off to college, I very much hoped to learn about "herps" in an academic context. There was still a rich array of naturalists on the faculty, and I managed through stealth and guile to take many of their courses. But I never got a herpetology class; my education in amphibians, such as it is, was self-won. I'll not forget the first time I went to hear an evening chorus at Carkeek Park, where a ravine runs down to Puget Sound in north Seattle. The raised voices of Pacific tree frogs erased all else in the night, even the sound of a passing train. (The creatures' sweet cacophony still graces the valley in which I live, from late February into April each spring.)

My younger brother, Howard Whetstone Pyle, whom we call Bud, took the nickname *Toad* in high school. Trying to remember why, as I first wrote this, I figured the name came about because he was a creative, nonconforming kid who identified with the self-contained individuality of toads, unconcerned with popularity. But when I ran this theory by him recently, Bud erupted with mirth and a definitive "Balderdash!"

"Why was it, then?" I asked.

"Too many Gomerisms!" he replied, laughing. Having also grown up with the name *Pyle,* I knew just what he meant. "Besides," he went on, "toads are cool beasts."

Back then, Bud made a sign for his room that read "Toad Abode," and we began exchanging toad amulets and figurines. (He made a classic for me in a high school art class: "Toad in a Stained-Glass Bog.") This led somehow to my collecting toad effigies—a hobby I've tried earnestly to limit ever since, with very little help from my friends. As an undergraduate struggling with chemistry when I wanted to be birding instead (usually succumbing, and consequently flunking chemistry), I developed a positive mania for toad collecting. Only now do I recognize it as a displacement activity. I wanted to be a naturalist, and the university wanted something else for me. One way to defuse the uncertainty and frustration was to put it all aside and go hunting for toads in strange places.

I would haunt junk shops, import stores, and Goodwill stores in search of toads of glass or porcelain, stone or metal, plastic, plaster, or wood. In those days, there was a St. Vincent de Paul thrift store down at the end of a dock on Lake Union, where rare artifacts might be unearthed from beneath heaps of stale discards and mounds of free shoes. Staffed by Dickensian figures in long, dirty coats, this warren was a vestige of an earlier age. Of course, we clothed and furnished our lives from these places, too, but I was always on the lookout for toads. I was not very discriminating. My sole criteria were diversity and that the objects be more toadlike than froglike (many finds, of course, showed hybrid characteristics). St. Vinnie's often came through.

Occasionally, I found a real prize: a big, fat sandy-backed Japanese toad with a toadlet on board, which still abides next to our stove; a glass *Rana amethysta* blown especially for me after I demonstrated the appropriate gestures and calls, on the isle of Murano, near Venice; a very old Chinese virility amulet prized from the depths of a true junk trove in the bowels of the Pike Place Market before its gentrification. This heavy brass toad crouches in a smooth form easily concealed in the folds of a gentleman's robe and bears on its ventral surface an immense erection. But the apotheosis of the toad collection was Dudley. Named for my robust high school discus-throwing partner, Dudley was a huge papier-mâché toad, correctly colored and proportioned, built by a friend named Marsha. Dudley occupied a significant part of several apartments and small houses until, during an extended trip of mine abroad, he hopped away into an unknown oblivion.

This peculiar compulsion aside, my life has been a vessel for many a flesh-and-blood toad since. There were the classic toads of English gardens, *Bufo bufo*, of which George Orwell writes so wonderfully, out among the snails (finally), the hedgehogs, and the gnomes. There was a tiny red-spotted toad negotiating the damp edge of a concrete cistern in the Chiricahua Mountains. Most recently, there came a toad of such stolid mass and magnitude that it reminded me of Dudley. We had just parked at a motel on the fringe of Tucson, Arizona, when observant Thea spotted a big lump beneath another car. "That's a toad!" she announced, "or else a big rock."

Toad it was. With imposing glands on its legs as well as viscid parotids, it did not invite handling, but we watched for some time as it hunted and snuffed crickets in the glow of the lamp standards. So large was this toad that we wondered whether it might be an introduced marine toad *(Bufo marinus)* from Brazil. Just days later, though, I read galleys of Ken Lamberton's *Wilderness and Razor Wire*, a remarkable account of a naturalist's life in prison and the weeds and creatures that kept him sane. One chapter eloquently tells the tales of the toads that haunt the edges of the prison yard—the spadefoots, after rain, and the Sonoran Desert toad (also called the Colorado River toad, *Bufo alvarius*). The latter is a lunker, and we had found this parking-lot monster on the very hems of the Sonoran Desert. I realized then that our dramatic encounter in the night had been with a native species—a true desert toad, surviving so far in spite of the worst Tucson could do. This lone toad, an asphalt epiphany, had called a single note in the night air—a note of hope for toads and their kin everywhere.

—ROBERT MICHAEL PYLE

Introduction

Amphibians have never succumbed easily to the logic of a simple either-or. In the drama of life on earth, they have for eons proven capable of playing multiple roles, emerging as the archetypal creatures of compromise. Faced with the primordial partition of the earth into water and land, they have been constitutionally disposed to strike a balance between the two. Aquatic and herbivorous as larvae, they graduate to a carnivorous terrestrial life. But their liberation from the liquid realm is never complete. Some measure of moisture is always necessary to meet the ever-present threat of desiccation, and a full retreat to water is typically required for the annual rite of breeding. This dual existence has rightly secured for these animals the designation *amphibious*—literally, living on both sides, leading double lives.*

Naturalists in every age have been intrigued by the lives of frogs, toads, and salamanders. They have noticed their general affinity for moist environments and have felt the dampness they carry with them on their skin. They could hardly fail to take note of the seasonal pronouncements of the more vocal members of the class, and they have been fascinated by the ensuing, and conveniently public, interplay of the sexes and emergence of the next generation. These investigators have seen amphibians in a variety of guises—as beings with magical powers or implicit moral lessons, as the products of spontaneous generation, as weather-wise heralds of the seasons, as evidence for evolution or material for biological experiments, and, most recently, as ecological barometers for the biosphere.

The writings of naturalists contained in this book thus reveal an abundance of interesting features of amphibian life. But they also, of course, reveal much about the writers' worldviews, particularly their sense of the relation between human beings and nature. The book's chronological arrangement enables the reader to see how these views have changed through the long course of Western history and to discover the niche amphibians have occupied in this cultural evolution.

From the earliest era of natural history in the Western world, students of nature have discerned patterns among the great diversity of creatures sharing

---

*For a summary of amphibian evolutionary history, see pages 168–169.

the earth with them, have divined an order in the apparent chaos and cacophony. Plato saw the immense variety of living things as the irrepressible creativity of a divine artisan striving for the full expression of eternal forms. Aristotle offered the more grounded suggestion that living things are arranged in an ascending series, a scale of perfection—plants, animals, human beings, and beyond—based on the degree of life or *psyche* they display. Thus, viviparous creatures were ranked higher than oviparous, and humans, the "rational animals," were considered the most perfect terrestrial beings. But nature will not go quietly into categories, and Aristotle would be the first to make the point. Intractable body parts and ingenious behaviors transgress the boundaries of our classes, as if to bridge the intervening gaps. This shading of one class into the next in the taxonomic twilight zones was evidence to Aristotle that the scale of creatures, though rough and complex, was continuous. And frogs, toads, and salamanders are good examples of this principle, defying any airtight grouping of animals by water and land.

Plato and Aristotle sowed here some potent seeds. Their ideas, known respectively as the principle of plenitude and the principle of continuity, became the main ingredients in the fertile notion of a grand scale of nature, a great chain of being, which was the almost unquestioned metaphysical map that gave naturalists their philosophical bearings for centuries. The golden chain, with innumerable links, hung from spiritual heights and reached into the basest material dregs of the earth. Humans were the middle link in this spiritual–material spectrum, compound beings—half spiritual, half material—charged with living a mixed existence that was the source of both heavenly opportunity and hellish anxiety. Sir Thomas Browne expressed it well, with a fitting analogy, in his *Religio Medici* in 1635, saying that we humans are

> that amphibious piece between a corporal and spiritual essence, that middle form that links those two together, and makes good the method of God and nature, that jumps not from extremes, but unites the incompatible distances by some middle and participating natures.[1]

Below us on the scale were the multitudes of animals, plants, and physical substances. But above us was even more—an almost infinite population of higher, more angelic beings inhabiting both this world and others. The human middle ground, though fully, and perhaps even uniquely, worthy of divine husbandry, was by no means seen as the prime real estate of the cosmos.

The lowly position of the earth in space was indeed rather analogous to the position of the humble amphibians among earthly creatures. Nevertheless, most people thought that the earth was where the action was. For in the very midst of the ubiquitous corruption and death on this terrestrial plane, pristine new life springs forth. And again, the amphibian's world is a sort of microcosm because out of the lowlands each year, a spirited, singing festival of resurgent life breaks out. The ancient Egyptians even dignified the phenomenon by incorporating the image of the frog in a hieroglyph to denote "repeating life"; later, the early Christians used it as a symbol of the Resurrection.

Because the conception of a great chain placed humans in many ways closer to the animals than to the angels, it could serve to deflate the billowing spinnaker of human pride. The message was, as expressed by Arthur O. Lovejoy, the great historian of the idea, that every link exists "not merely and not primarily for the benefit of any other link, but for its own sake, or more precisely, for the sake of the completeness of the series of forms, the realization of which was the chief object of God in creating the world."[2] So every link was golden, every creature good. English philosopher Henry More sermonized in 1653 that we should not consider the fineness of design evident even in frogs to be an affront to our moral sensibilities, as a waste of the Creator's talent:

> For this only comes out of pride and ignorance, or a haughty presumption that because we are encouraged to believe that in some sense all things are made for Man, that therefore they are not made at all for themselves. But he that pronounces thus, is ignorant of the nature of God and the knowledge of things. For if a good man be merciful to his beast, then surely a good God is bountiful and benign, and takes pleasure that all his creatures enjoy themselves that have life and sense and are capable of any enjoyment.[3]

Henry More, like Plato before him, imagined that the symphony of life on earth was orchestrated by the *anima mundi,* the soul of the world, which informs each organism, even "those little soules" that swarm over the marsh or sit at its edge. If nature, in both particulars and grand plan, is divine artistry, then even the lowly and despised creatures possess a certain elegance. Thomas Browne opined, "I cannot tell by what Logick we call a *Toad,* a *Bear,* or an *Elephant* ugly, they being created in those outward shapes and figures which best express the actions of their inward forms."[4]

Naturalists of all eras have, of course, noticed that organisms normally arise from parents of the same type—that like begets like—and that this process of generation is the main route by which "inward forms" propagate additional organisms with similar "outward shapes." However, from ancient times until at least the seventeenth century, most biologists allowed room in their philosophies of nature for some form of spontaneous generation. Salamanders, whose reproductive activity through the female assumption of spermatophores deposited in pond or field is somewhat less obvious than that of frogs, were for centuries seen as somewhat of a classic anomaly. They were considered not only to be so cold that they could survive or even extinguish flames but also perhaps even to be spontaneously engendered in fire, as Pliny surmised, "from some hidden and secret source."

If the earth's natural entities—from minerals to mussels to monkeys to me—are truly ordered in a continuous series, with each life-form shading insensibly into the next, then the prospects for spontaneous generation at the lowest levels of the scale are perhaps rather bright. But during the seventeenth century, leading naturalists placed the emphasis much more squarely on discreteness than on continuity, on particulate species rather than on a seamless wave of natural forms. The idea of fixed species was then linked to the process of normal generation, which was thought to keep them fixed. In 1686, English naturalist John Ray defined species in his *General History of Plants* in terms of "distinguishing features that perpetuate themselves in propagation from seed," emphasizing that "one species never springs from the seed of another." He thus took pains to disprove the common belief that frogs could be engendered in clouds and fall with the rain. To accept such an "utterly false and ridiculous" notion would be to allow the unwelcome element of chance into a divinely ordered, strictly speciated world.

Dutchman Jan Swammerdam saw the situation similarly and found in frog eggs additional marks of providence. He opposed not only the idea of spontaneous generation but also that of metamorphosis because of their presumably atheistic implications of irregularity in nature. He identified the black sphere at the center of a frog's egg as a little frog and argued that the speck did not change in essence as it grew; it merely unfolded mechanically its secret structures, "displaying itself under a variety of forms," as it increased materially. There was thus nothing new under even the warm spring sun, which meant that the seed of every living thing must have been in existence from the beginning of time. When Swammerdam shared his amphibian findings with a philosopher, the man proposed that such preexisting seeds could explain how the sin of Adam had infected humankind and that the species would end when Eve's eggs were used up.

The question of the day regarding species was whether they were the product of the Divine Mind or of merely human minds. The argument turned on whether species were believed to be real, objective features of nature or just artificial but convenient ways of organizing and conversing about the natural world. The great Swedish taxonomist Carolus Linnaeus was a believer, saying in several of his works that "there are as many species as the Creator produced different forms in the beginning." Each of these had its part to play in the natural economy of Linnaeus, but he, unlike Thomas Browne a century earlier, did not find beauty in *every* face, least of all in those of amphibians, a class composed of "most terrible and vile animals." He continued his unflattering depiction by saying that "most amphibians are rough, with a cold body, a ghastly color, cartilaginous skeleton, foul skin, fierce face, a meditative gaze, a foul odor, a harsh call, a squalid habitat, and terrible venom. Their Author has not, therefore, done much boasting on their account." Yet lovely or not, they were distinctive and diverse characters in the language of nature, and made their meditative presence felt with feet firmly planted in the real world. Or in two worlds, because, as Linnaeus noted, "a polymorphous nature has bestowed a double life on most of these amphibians."[5]

Now, the best of all possible worlds for those who envisioned a great chain of being would have been one with the fullest spectrum of life, with no missing links. Most naturalists thought that the actual world was indeed just such a place and that any apparent gaps in the order of creatures were an indication of the limits of human knowledge, not a sign of discontinuity in nature. Here, the metaphysical hypothesis launched a scientific research program—the search for missing links. It was a cause the Royal Society of London took to heart from its founding in 1660, with the energizing conviction that "this is the highest pitch of humane reason: to follow all the links of this chain, till all their secrets are open to our minds; and their works advanced or imitated by our hands. This is truly to command the world."[6] During the eighteenth century, botanical and zoological expeditions were mounted far and wide and the number of known species rose dramatically, straining the taxonomic capacity of a linear series. Linnaeus thought he had been afforded a glimpse of the divine logic linking all creatures, but to him the plan looked more like an extensive map of interlocking territories than a single, linear chain. He developed the familiar binomial (genus-species) approach to chart this increasingly complex biological terrain.

Amphibians and reptiles were, of course, an important part of this great migration of creatures into classification schemes and specimen boxes. They became especially numerous at the Museum of Natural History in Paris, which was then the largest research institution in the world and had the best herpetological collection. Many of the greatest zoologists of the age staked their claims to fame at the museum, and two of them succeeded in sorting reptiles and amphibians into separate classes. Whereas Linnaeus's "Amphibia" encompassed both amphibians and reptiles (as well as cartilaginous fishes), in 1800 Alexandre Brongniart made a close anatomical study of herpetofauna, a method inspired by Georges Cuvier, and noticed that one group of reptiles—the batrachians—was significantly different from the rest. A few years later, Pierre-André Latreille, a colleague of Brongniart, magnified the distinction by placing the batrachians in a separate class, the amphibians.

By the time Charles Darwin boarded the HMS *Beagle* in 1831, belief in the objective reality and immutability of species was being severely tested. The critical factor that eventually settled the case for Darwin was his study of the geographical distribution of different species. On his voyage through the islands of the South Pacific, he noticed that these isolated outcrops had their own rather distinctive plant and animal populations, which, however, resembled those of the nearest landmass. He theorized that the intervening expanses of sea acted as barriers, creating separate populations—those of reptiles and birds were especially significant—and environments in which organisms, though originally identical, gradually evolved into separate, reproductively isolated species. He also was struck by the absence of various types of animals, and he specifically mentioned amphibians, which he suggested would have fit quite well into the island habitat. He again looked to the ocean to explain this fact, noting that salt water would have spelled doom for any amphibian eggs drifting islandward from mainland shores. "But why, on the theory of creation, they should not have been created there," he remarked in *The Origin of Species*, "it would be very difficult to explain."

The idea that the action of merely physical forces, rather than the artistic impulses of God, brought forth distinct species diffused the notion of plenitude—that the fullness of forms is a divine necessity—and had the effect of leveling the great chain of being. With the advent of Darwinism, many naturalists thought less of a linear hierarchy of beings animated by spiritual design and more of a horizontal distribution of organisms assailed by material imperatives. In the course of the nineteenth century, the image of the chain of being was largely replaced by the more expansive and dynamic one of the tree of life. Darwin's version of this tree had the curious feature of competition among its growing twigs in the struggle for survival. But it had the great virtue of emphasizing that all the twigs and branches—amphibians, humans, and all—share the roots, and the struggle, which some branches of humanity have been prone to forget.

Whether one wrapped the world in a golden chain or cultivated a tree of life, it was possible to see oneself as a member of a larger community of creatures, and even to muster a certain reverence for all life. Possible, but perhaps not probable. More common was the traditional perspective, amplified by Francis Bacon and codified in the research program of the Royal Society, that tracing each link in the chain meant progress in "commanding the world," gaining dominion over an earth made for human use. Two centuries later, Thomas Henry Huxley relished his role of "Darwin's bulldog" but lacked the master's creaturely sympathies. His confrontations on behalf of Darwinism were fed by his sense of a larger battle between humanity and nature, a just war waged in the hope of transforming the earth and lower animals into, in Huxley's favorite image, "a true Garden of Eden, in which all things should work together towards the well-being of the gardeners."[7]

The widespread success of the Baconian and Huxlean plan for transforming the earth has resulted in a world in which the well-being of both the gardeners and the garden is in question. The divorce of humanity from the natural order—the distressing either-or at the core of much modern life—has resulted in a broken household with strained relations between its alienated, and increasingly absent, inhabitants. And for more than two hundred years, a growing company of nature writers has been trying to mediate the dispute. From English curate Gilbert White's affecting eighteenth-century depictions of pastoral harmony at Selborne through nineteenth-century Romantic portraits of animated nature on both sides of the Atlantic Ocean to a host of twentieth-century missives fashioning lessons from the land, these writers' works have borne witness to the inescapable fact of human membership in the family of life, and the value—indeed, the rising necessity—of *conscious* membership.

One of the curious features of the impressive edifice of modern biological science is its lack of a precise estimate of the number of plant and animal species on earth. Educated guesses range from 5 million to 30 million. Of these, about 1.5 million have been formally described. Among these known organisms, the frog is perhaps known in the most exhaustive detail, having been the subject of myriad physiological experiments and untold field studies. Many toads and salamanders are also, of course, relatively well known, in part because of their rather easy accessibility. Their worlds are often next door to ours. At present, the number of described species of amphibians is very near 5,000, with a little more than 4,000 being frogs and toads and the remainder being salamanders and the less familiar, wormlike caecilians.

Considering this depth of knowledge about amphibians, it is surprising that biologists in recent years have been discovering so many new species. In fact, this class of organisms has one of the highest proportional rates of description of new species—about 25 percent over the past fifteen years. This mushrooming of zoological knowledge, somewhat reminiscent of the great eighteenth-century growth, has resulted both from the discovery of previously unknown species in the field (even a field as nearby as the San Gabriel Mountains, just east of Los Angeles, where a new species of salamander was found in 1996) and from the application of molecular taxonomic tools to members of a presumably single species only to find multiple genetically distinct species within it. Another factor in the appearance of new species, and another echo of the eighteenth century, is the current debate over the definition of species, with the traditional reproductive isolation concept being challenged by an emphasis on evolutionary history and independence. The dramatic rise in species from these various sources currently shows no sign of slowing.

But at the same time that this multiplication of known species of amphibians is taking place, there is also occurring on our planet a great and troubling process of subtraction. Herpetologists began to notice scattered evidence of the decline or disappearance of numerous amphibian species more than twenty years ago and recognized it as a global phenomenon in 1990. Since then, this multifaceted ecological puzzle, compounded by the more recent issue of amphibian deformities, has engendered headlines around the world and headaches among many herpetologists. The declines and deformities have attracted great interest in part because amphibians are considered by many biologists to be particularly sensitive biological indicators, early warning signals of the declining health of the larger environment—canaries in the ecological coal mine.

The culprits appear to be many and varied, and there is some evidence of collusion. Some are clearly anthropogenic—particularly habitat destruction, pollution, and ozone depletion—whereas others seem more biogenic—a deadly fungus and a deforming parasite. But the scene is riddled with complications—different factors at work in diverse locations, interacting causes at single sites, and a lack of long-term population statistics—and investigators have a good deal of counting and sifting of evidence still to do.

But preservation efforts cannot wait for a final verdict or, more likely, verdicts. The concern with preserving amphibian diversity is, of course, merely an aspect of the larger concern for biodiversity as a whole. And the reasons for such preservation fall into familiar categories. Perhaps most obvious are the anthropocentric arguments. Frog legs have landed on dinner plates for centuries, having made their fateful leap into haute cuisine as early as the seventeenth century. Frogs also provide easy but wondrous biology lessons; millions of schoolchildren have cut their biological teeth on the body of a leopard frog or bullfrog laid out revealingly in a dissecting tray or have been initiated into nature's mysteries by watching the conspicuous pageant of metamorphosis. More seasoned inquirers have found amphibians to be ideal organisms for a range of studies that have advanced scientific knowledge on many fronts and have in the process acquired two Nobel Prizes. A variety of pharmaceuticals have been derived from amphibians, particularly from some of the more exotic species, whose skin secretions offer powerful painkillers and possible treatments for Alzheimer's disease and peptic ulcer, among other ailments. Amphibians also have long contributed to the human aesthetic experience, particularly by means of their vivid coloration, gilded eyes, and seasonal serenades.

But the human-centered perspective has its limits. It is often quite possible to find substitutes for the particular benefits derived from amphibians, and in light of current ecological concerns, such substitutes should be avidly pursued. Moreover, although the rationality of this point of view appeals to the head, it perhaps fails to reach the heart of our concern with biodiversity—our sense of the intrinsic value of life-forms other than ourselves. Presenting biocentric rather than anthropocentric arguments requires more philosophical groundwork, more tangling with deep-rooted beliefs, the possibilities of this perspective being less obvious to most people. Space here does not allow a full tilling; however, the image of the web of life is a good place to start. We all realize to some degree that our life is linked to the life around us, that our very survival depends on these relations—to family and friends, to society, to a whole range of technological devices, and, very basically, to food, clothing, and shelter. Indeed, our existence consists in being the focus of these relations, a vital center in the web of life. A web, of course, has many nodes, and it is perhaps not too difficult to imagine each one—the frog at water's edge, the salamander under a log—as, like ourselves, a focus of relations, a vital center of life, and thus worthy, like us and our kin, of careful consideration. It is an argument with a good deal of philosophical weight on its side but with the balance of cultural sentiment and tradition aligned against it because it requires us to surrender the comfortable dualism that fashions human beings as the only true subjects on earth—the only beings with an "inside"—and places us above and beyond insensible nature.

A biocentric view can teach the inherent value of individual amphibians or other organisms, but it does not answer the question of why *many* kinds of lifeforms are better than a few. For that, a larger vision is needed. Ecological science reminds us that the web of life is a dynamic network of energy flows and that amphibians occupy critical niches in this system. The total amphibian biomass per acre in many temperate and tropical habitats, for example, often exceeds that of reptiles, birds, or mammals. And amphibians are extremely efficient conveyors of energy. Tadpoles graze on algae and thus pass plant energy on to their predators, and in many damp locales, adult amphibians are the leading vertebrate predators on invertebrates. As they in turn become prey for higher vertebrates, the buzzing invertebrate energy rises higher in the food chain. Further, an abundance of ecological research indicates that the greater the diversity of organisms, the greater the integrity and resilience of an ecosystem as a whole. The same principle seems to hold at the planetary level, where, according to Gaia theory, diversity of ecosystems translates into greater geophysiological health of the earth.

If the earth as a whole does have a unity, an integrity that makes it more than the sum of its parts, then each of these parts—each species of frog, toad, and salamander—makes a contribution to that wholeness. With this perspective, we are in the neighborhood of the *anima mundi* of Plato and Henry More, and close to the latter's view that all creatures with life and sense are capable of some measure of enjoyment. The spark of amphibian inner life burns with distinctive hues in the larger spectrum of life on earth, expressing the beauty of its inward form, as Aristotle and Thomas Browne intoned, and the loss of any species darkens and diminishes that larger life. If the world has a soul, which many thinkers for most of Western history have considered it reasonable to believe, are the jeweled eyes of amphibians perhaps humble ancient windows of that soul?

A biocentric perspective is, of course, rather far from the hearts of the general public. Certainly such views have gained popular and even some legal ground in recent decades. But the dualism that fractures human beings and the natural world is still second nature to so many of us. If we continue to succumb to the perverse logic of this tragic either-or, many species of amphibians, which for ages have survived in the face of other challenging dualities by responding with a creative "both-and," may be defeated by the severity of this one. Amphibians today would perhaps echo, if they could, Aristophanes' ancient chorus of frogs, which, when Bacchus forbade them to continue their singing, replied:

> That would be severe indeed;
> Arbitrary, bold, and rash—
> Brekeke-kesh, koash, koash.[8]

The selections included in this anthology are drawn from the entire Western natural history tradition. I had several criteria in mind when choosing particular texts. In the first three parts, I focused on those individuals who have been influential in shaping that tradition. And I included classic texts that best illustrate central themes in the changing understanding of amphibians in particular and of the natural world in general. For the fourth part, I chose engaging essays by leading twentieth-century nature writers portraying a variety of amphibians in diverse terrains. In both of the nature writing sections—part II and part IV—I limited the selections to the Anglo-American tradition, both to keep the project manageable and because, for various historical reasons, nature writing in this tradition has developed into an especially rich and influential literary form. Finally, the selections in part V cover the various aspects of, and research on, the problems of amphibian decline and deformities. Throughout the collection, I kept readability in mind so that the book might not only inform but also entertain and even inspire.

Part I

Interpreting the Cosmos
Early Naturalists

In the ancient Mediterranean world, the springtime activity in the swamps was pregnant with cosmic significance. When the croaking of frogs entered the ears of Aristotle, he recognized it as an element in the grand cycle of the seasons, part of the periodicities of nature that formed the heart of the widespread sense of a *kosmos*—an orderly, harmonious universe. Aristotle, Pliny the Elder, and other early naturalists also noticed that this song of the swamps was not just idle chatter. It was a love song, and it led to things—to nocturnal liaisons, and to tadpoles. In a finely designed cosmos, they thought, nature does nothing in vain; each part and process has a purpose, every species a proper place. Aristotle and his followers thus spawned a style of natural history in which all aspects of nature, including animals of the swamps, are worthy of close study because "in not one of them is Nature or Beauty lacking."[1] The writers represented in this first part bear witness to the diversity and grandeur of the natural world and to the pathbreaking challenge they faced in trying to comprehend it.

If natural things did not lack beauty, it was because they seemed, by and large, to become what they were apparently meant to be. Early natural philosophers thus found much they could rely on in the sequences of natural events, such as the annual resurrection of amphibian life and the progression from croaking frogs through tadpoles to future frogs. But perfect lawfulness could not be found in every corner of the kingdom, and the phenomenon of generation is a prime example of both the exception and the rule. All students of nature have, of course, recognized that the generation of offspring typically consists in making a reproduction, or at least a resemblance, of the parents. But in earlier centuries, generation of organisms from dissimilar substances was also considered possible, as is evident in the selections from Aristotle, Pliny, and Edward Topsell that follow. Most early naturalists entertained this possibility at least in part because of the diversity of influences they saw impinging on natural things. In addition to the various terrestrial conditions familiar to modern minds, there was a whole range of celestial influences streaming in from the stars. This astrological outlook was especially strong during the Middle Ages and the Renaissance and formed what historian Clarence Glacken has called a "cosmic environmentalism," in which each organism and location on the earth was seen as subject to a certain set of stars above. These astral energies knit together the cosmos and formed sympathetic relations between many of its myriad parts, especially those with apparently similar forms or functions. Thus, the tongue of a talkative frog was employed to affect speech in other organisms, and bulging amphibian eyes were used to remedy visual afflictions. Such magical uses of frogs and toads were reported, with various assessments of their efficacy, by Pliny, Albert the Great, and Topsell.

In a world with a presumed coordination between the earthly and heavenly realms, mundane plants and animals were endowed with transcendent meanings. As far back as the second century, the anonymous Christian author of the *Physiologus* assumed that frogs and salamanders were interesting not primarily in themselves but for the human lessons they afforded. This work, which apparently drew on the writings of Pliny and other ancient sources, became the model for innumerable moralizing medieval bestiaries. Each plant was a letter in an intricate natural alphabet, each animal a symbol in a divine language of such depth that everything was rich with hidden meanings. From this perspective, a frog was never really just a frog, and to gain a merely anatomical, physiological, or even ecological understanding of it was merely to scratch the surface. The frog was stitched into the warp and woof of the cosmos, and apprehending and aligning oneself with the patterns in this encompassing web of life was a matter of supreme significance.

This transcendental ecology was practiced for centuries in various forms, with one of the foremost Renaissance exemplars being Topsell's *History of Four-Footed Beasts and Serpents* (1608), which is based in large part on Swiss naturalist Conrad Gesner's monumental *History of Animals* (1551–1587). Although these authors spared their readers the moral lessons of the medieval bestiaries, they shared with earlier writers a conviction that natural objects were symbols or emblems of larger realities and thus that true knowledge of things required awareness of a vast range of earthly and heavenly relations. With thousands of published pages to their credit, they obviously were intent on explicating these links in almost exhaustive detail.

By the time Topsell's book was published, in the early seventeenth century, however, many natural philosophers had grown weary of the emblematic worldview and were suggesting that the language of nature was actually much simpler and less equivocal. Galileo issued an early challenge by declaring in 1623 that the book of nature "is written in the language of mathematics."[2] The rather messy stuff of biology did not lend itself to the rule of measure as easily as did falling spheres and rays of light, but a growing number of naturalists were also certain that if viewed from the proper angle, the logic of life on earth would become clearer and more consistent. This meant casting off the aura of magic and symbolic associations surrounding natural things, and with it the burden of ancient authorities and received truth, and putting some things to the test. Biologist François Jacob has recently given a vivid description of this shakedown:

> ... living bodies were scraped clean, so to speak. They shook off all their crust of analogies, resemblances, and signs, to appear in all the nakedness of their true outer shape.... What was read or related no longer carried the weight of what was seen.... Instead of a contemplation, an exegesis or a riddle, natural science becomes a decoding.... What counted was not so much the code used by God for creating nature as that sought by man for understanding it.[3]

Sir Thomas Browne was one of the first figures to breathe this new spirit into natural history, and his main work on the subject, the 1646 *Pseudodoxia Epidemica,* carried the subtitle *Enquiries into Very Many Received Tenents and Commonly Presumed Truths.* Instead of merely speculating or consulting revered authorities to decide some persistent questions of natural history—such as those pertaining to the generation of frogs, the salamander's immunity to fire, and the long-supposed antipathy between spiders and toads—he boldly set out to give these things a try. Dutch naturalist Jan Swammerdam and Englishman John Ray also avidly pursued this path toward a purified modern science based on systematic observation and experiment. For them, the emergence and unfolding of tadpoles into frogs offered evidence against the unlawful, and thus atheistic, notion of spontaneous generation and proof of the unfailing wisdom and providence of the Creator in a clockwork cosmos.

Aristotle

From *Historia Animalium* (*Inquiry Concerning Animals*) (fourth century B.C.E.)

... Voice and sound are different from one another; and language differs from voice and sound. The fact is that no animal can give utterance to voice except by the action of the pharynx, and consequently such animals as are devoid of lung have no voice; and language is the articulation of vocal sounds by the instrumentality of the tongue. Thus, the voice and larynx can emit vocal or vowel sounds; non-vocal or consonantal sounds are made by the tongue and the lips; and out of these vocal and non-vocal sounds language is composed. Consequently, animals that have no tongue at all or that have a tongue not freely detached, have neither voice nor language; although, by the way, they may be enabled to make noises or sounds by other organs than the tongue. ...

Of animals which are furnished with tongue and lung, the oviparous quadrupeds produce a voice, but a feeble one; in some cases, a shrill piping sound, like the serpent; in others, a thin faint cry; in others, a low hiss, like the tortoise. The formation of the tongue in the frog is exceptional. The front part of the tongue, which in other animals is detached, is tightly fixed in the frog as it is in all fishes; but the part towards the pharynx is freely detached, and may, so to speak, be spat outwards, and it is with this that it makes its peculiar croak. The croaking that goes on in the marsh is the call of the males to the females at rutting time; and, by the way, all animals have a special cry for the like end at the like season, as is observed in the case of goats, swine, and sheep. The bull-frog makes its croaking noise by putting its under jaw on a level with the surface of the water and extending its upper jaw to its utmost capacity. The tension is so great that the upper jaw becomes transparent, and the animal's eyes shine through the jaw like lamps; for, by the way, the commerce of the sexes takes place usually in the night time.... (Book 4, chap. 9)

Now there is one property that animals are found to have in common with plants. For some plants are generated from the seed of plants, whilst other plants are self-generated through the formation of some elemental principle similar to a seed; and of these latter plants some derive their nutriment from the ground, whilst others grow inside other plants, as is mentioned, by the way, in my treatise on Botany. So with animals, some spring from parent animals

From *Historia Animalium (Inquiry Concerning Animals)* (fourth century B.C.E.)

according to their kind, whilst others grow spontaneously and not from kindred stock; and of these instances of spontaneous generation some come from putrefying earth or vegetable matter, as is the case with a number of insects, while others are spontaneously generated in the inside of animals out of the secretions of their several organs. . . .

And, by the way, living animals are found in substances that are usually supposed to be incapable of putrefaction; for instance, worms are found in long-lying snow; and snow of this description gets reddish in colour, and the grub that is engendered in it is red, as might have been expected, and it is also hairy. The grubs found in the snows of Media are large and white; and all such grubs are little disposed to motion. In Cyprus, in places where copper-ore is smelted, with heaps of the ore piled on day after day, an animal is engendered in the fire, somewhat larger than a bluebottle fly, furnished with wings, which can hop or crawl through the fire. And the grubs and these latter animals perish when you keep the one away from the fire and the other from the snow. Now the salamander is a clear case in point, to show us that animals do actually exist that fire cannot destroy; for this creature, so the story goes, not only walks through the fire but puts it out in doing so. (Book 5, chaps. 1 and 19)

Figure 1
Salamander in fire. (From Michael Maier, *Atalanta Fugiens,* 1617.)

Pliny the Elder

From *Natural History* (first century C.E.)

The curiosity and wonder of mankind does not allow us to postpone the consideration of these animals' method of reproduction.... Frogs cover the female, the male grasping her shoulder-blades with his fore-feet and her buttocks with his hind feet. They spawn very small lumps of dark flesh that are called tadpoles, possessing only eyes and a tail; but soon feet are formed by the tail dividing into two hind legs. And strange to say, after six months of life they melt invisibly back into mud, and again in the waters of springtime are reborn what they were before, equally owing to some hidden principle of nature, as it occurs every year.... (Book 9, sec. 74)

We have it from many authorities that a snake may be born from the spinal marrow of a human being. For a number of animals spring from some hidden and secret source, even in the quadruped class, for instance salamanders, a creature shaped like a lizard, covered with spots, never appearing except in great rains and disappearing in fine weather. It is so chilly that it puts out fire by its contact, in the same way as ice does. It vomits from its mouth a milky slaver, one touch of which on any part of the human body causes all the hair to drop off, and the portion touched changes its colour and breaks out in a tetter.

Consequently some creatures are born from parents that themselves were not born and were without any similar origin, like the ones mentioned above and all those that are produced by the spring and a fixed season of the year. Some of these are infertile, for instance the salamander, and in these there is no male or female, as also there is no sex in eels and all the species that are neither viviparous nor oviparous; also oysters and the other creatures clinging to the bottom of shallow water or to rocks are neuters. But self-generated creatures if divided into males and females do produce an offspring by coupling, but it is imperfect and unlike the parent and not productive in its turn: for instance flies produce maggots. This is shown more clearly by the nature of the creatures called insects, all of which are difficult to describe and must be discussed in a work devoted specially to them.... (Book 10, secs. 86–87)

In frogs the tip of the tongue is attached but the inner part is loose from the throat; it is with this that the males croak, at the time when they are called croakers; this happens at a fixed season, when they are calling the females to mate. In this process they just drop the lower lip and take into the throat a moderate amount of water and let the tongue vibrate in it so as to make it undulate, and a croaking sound is forced out; during this the curves of the cheeks are distended and become transparent, and the eyes stand out blazing with the exertion.... (Book 11, sec. 65)

I will arrange water creatures according to diseases, not that I do not know that a complete account of each living thing is more attractive and more wonderful, but it is more useful to mankind to have remedies grouped into classes, since they vary with individuals, and are more easily found in one place than in another....

A decoction of sea frogs boiled down in wine and vinegar is drunk to counteract poisons, also that of the bramble toad and salamander; if the flesh of river frogs is eaten, or the broth drunk after boiling them down, it counteracts the poison of the sea-hare, of the snakes mentioned above, and of scorpions if wine is used in the preparation. Democritus indeed tells us that if the tongue, with no other flesh adhering, is extracted from a living frog, and after the frog has been set free into water, placed over the beating heart of a sleeping woman, she will give true answers to all questions....

The right eye of a frog hung round the neck in a piece of undyed cloth cures ophthalmia in the right eye; the left eye similarly tied cures ophthalmia in the left. But if the frog's eyes are gouged out when the moon is in conjunction, and worn similarly by the patient, enclosed in an eggshell, it will also cure albugo. The rest of the flesh, if applied, quickly takes away bruises. An amulet of crabs' eyes also, worn on the neck, are said to cure ophthalmia. There is a small frog, found living especially in reed-beds and grasses, deaf, without a croak, and green, which, if it by chance is swallowed, swells up the bellies of oxen. They say that the fluid of its body, scraped off with a spatula and applied to the eyes, improves vision. The flesh by itself is placed over painful eyes. Some put together into a new earthen jar fifteen frogs, piercing them with rushes; to the fluid that thus exudes they add the gum of the white vine, and so treat eyelids; superfluous hairs are plucked out, and the mixture dropped with a needle into the holes made by the plucked-out hairs. Meges used to make a depilatory for the eyelids by killing frogs in vinegar and letting them putrefy; for this purpose he used the many spotted frogs that breed in the autumn rains.... (Book 32, secs. 15, 18, 24)

Anonymous

From the *Physiologus* (ca. second century C.E.)

"The Frog"

There is a frog called the Land-frog. Physiologus relates of him that he can sustain the heat and the glow of the sun, but when the rain touches him he dies. But the Water-frog, when it rises out of the water so that the sun's ray touches it, immediately it dives under the water again.

Now the Land-frog resembles good men who support the heat of temptation and yet, when the cold winter of persecution touches them, then their virtue perishes. But the children of this world are Water-frogs. When the sun of temptation touches them, then they dive deeper in their former wanton passions.

[From Ponce de Leon] The Land-frog sustains the warmth of the sun; he also endures cold, rain, wind, and winter-storms. But the Water-frog can bear none of these: at the beginning of the winter he sinks himself in the depths; but, when the sun shines brightly, he comes out of the water and basks in the warmth—yet, when the heat of the sun increases, he can no longer endure it, and dives again into the depths.

So also those monks who spend their time in idleness cannot endure hunger, thirst, nakedness, abstinence, and lying on the hard ground. But those who do not yield to laziness, fast willingly, and endure hardships.

"The Salamander"

It is written: "When thou passest through fire the flame shall not burn thee." (Is. 43:2)

Physiologus relates of the Salamander that, when it enters the furnace, it puts out the flame; and, when it enters the oven, it makes the whole bath cold.

If now the Salamander extinguishes fire by its own attribute, how much more should the righteous by their own godly behaviour extinguish the fire, as they stopped the mouths of lions, and as the three men cast in the fiery furnace suffered no harm and indeed put out the fires of the furnace?

Figure 2
"Water frogs" and "land frog." (From *Physiologus*, 1587 G. Ponce de Leon edition.)

Albert the Great

From *De Animalibus (Man and the Beasts)* (ca. 1250)

Bufo (toad) is a four-legged worm [vermis] shaped like a frog but having a remarkably thick skin of ashen-grey color. In fact, the very thickness and resilience of its skin allows it to escape the harm of a direct blow. Its bite is poisonous, like a serpent of the second category. Since it makes its home underground, it lives on earthly humor and plants, and occasionally eats earthworms. But, the toad itself is eaten by the mole.

They claim the toad consumes no earthly moisture during one day other than what it could grasp in one hand on a single occasion, because it fears the whole earth would not be sufficient for its needs (if it gave way to unbridled eating). But, this story stems from folklore and is unsupported by factual evidence. When a toad has been soaked in water and then covered with salt, it first swells and next bursts and finally disintegrates down to the bones.

There is a particular species of toads that are called "horned" [cornuti] from the hornlike sound of their voices. These have a brownish-gray color with a yellow belly and tend to reside in stagnant marshes where they emit their calls, one to another. It is said these toads are heard nowhere outside the boundaries of France, but in my travels throughout the whole of Germany I have heard them calling very loudly, so the story is untrue.

Borax (midwife toad) is a brown species of toad, very large in size and sometimes attaining a cubit in size. It lives in warm countries and at times is accustomed to carrying its offspring on its back. This species of toad customarily bears a stone in its forehead, for which it is killed; this stone comes in various colors; sometimes a white stone is recovered from the toad's head, and this is thought to be the most valuable; another brownish-black type is found, and this is valuable when it has a yellow spot in the center; sometimes it is found to have a sickly color; during my lifetime a wholly green toad-stone was found; and occasionally the outline of a toad is depicted on the surface of the stone.

Like the snake, the midwife toad is a natural enemy of the spider. A spider, descending from above and hovering at the end of a filament, stings the brain of either one of these enemies. The toad, enraged by the bite, swells up and sometimes bursts asunder from the effects of the spider's poison. The midwife toad rarely appears during the day, except in very isolated areas, and it emerges from the ground only during a rainstorm. But, it sometimes comes out at night and then freely traverses the footpaths used by man. It detests rue and the flowers of the grapevine, both of which are employed to repel it along with other poisonous creatures.

Rana (frog) is a four-legged worm shaped like a toad but lacking its poison. Its long hindlegs and short forelegs have elongated digits with interdigital webs for swimming. Its tongue clings to the palate, giving its voice a croaking sound that emerges from the gullet to the mouth; since its vocal spirit or breath does not proceed in a straight line due to the interference of the tongue, it produces two inflated pouches on each side of the mouth. When the frog croaks, it holds its lower lip at the surface of the water and raises its upper lip above the water line. Only the male makes the croaking sound when he calls to the female. There is a popular opinion that in August the frog holds its lips so tightly closed they cannot be pried open, even with a metal instrument. When a frog is frightened by being removed from the water, it emits a weak cry like a scared mouse.

The frog mates in the springtime and lays many eggs in the water during the spring of the following year. In the center of these eggs a tiny frog lies hidden. When the young emerge from the eggs, they have a large head attached to an abdomen that tapers at the rear to a tail equipped with fins for swimming. After the month of May the tail is shed and four legs sprout from the body. During the winter the frog hibernates outside the water in warm crevices and occasionally under the mud in bodies of water that are relatively warm in winter. Then, in the spring it returns to the water. Sometimes when it begins to feel the chill of autumn, it enters human abodes where it has been known to hop onto a man's lap or abdomen. Some people say that if the tongue of a croaking aquatic frog is placed on the head of a sleeper, he will talk in his sleep and reveal secrets.

A certain frog is called "rubeta" or "rubetum" because it usually sits on a bramblebush [rubo] or thicket of reeds [harundineto]. This frog leads an amphibious life, spending equal time on land and in the water; it may be classified as poisonous, though its venom is scanty. Some marvelous stories are related about this frog. For example, a hook baited with this species lures and attracts the purple murices of the sea. If one places an ossicle obtained from the right side of this frog into a vessel of boiling water, the boiling will cease and not resume until the little bone is removed from the water. The charred remains of this frog, like the ashes of a sea hare [leporis marini], is an antidote against its own poison.

A tiny voiceless frog that lives among the reeds is sometimes inadvertently imbibed by cattle drinking the nearby water; the cows' bellies become so distended that they seem ready to burst.

There is also a green frog that climbs trees and foretells rainstorms by its peeping, but remains silent at other times. According to common belief, the incessant barking of a dog can be stopped by tossing one of these frogs into its mouth. (Book 26)

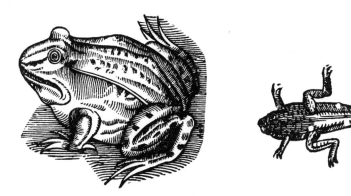

Figure 3
Frog and tadpole. (From Edward Topsell, *The History of Four-Footed Beasts and Serpents,* 1658 edition.)

Edward Topsell

From *The History of Four-Footed Beasts and Serpents* (1608)

... In the next place we are to consider the diversity and kinds of frogs, as they are distinguished by the place of their abode: for the greatest difference is drawn from thence; some of them therefore are water-frogs, and some are frogs of the land: the water-frogs live both in the water and on the land, in marshes, standing pools, running streams, and banks of rivers, but never in the sea; and therefore *Rana marina* is to be understood of a fish, and not a frog, as Massarius hath learnedly proved against Marcellus. The frogs of the land are distinguished by their living in gardens, in meadows, in hollow rocks, and among fruits: all which several differences shall be afterward expressed, with their pictures in their due places: here only I purpose to talk of the vulgar and common frog, whose picture with her young one is formerly expressed [see Figure 3 above]. Beside these differ in generation: for some of them are engendered by carnal copulation, and of the slime and rottenness of the earth. Some are of a green colour, and those are eaten in Germany and in Flanders; some again are yellow, and some of an ash-colour, some spotted and some black, and in outward form and fashion they resemble a toad, but yet they are without venom, and the female is always greater [than] the male: when the Egyptians will signify an impudent man, and yet one that hath a good quick sight, they picture a frog, because he liveth continually in the mire, and hath no blood in his body, but about his eyes. . . .

Sometimes they enter into their holes in autumn before winter, and in the springtime come out again. When with their croaking voices the male provoketh the female to carnal copulation, which he performeth not by the mouth (as some have thought) but by covering her back; the instrument of generation meeting in the hinder parts, and this they perform in the night season, nature teaching them the modesty or shamefastness of this action: and besides in that time they have more security to give themselves to mutual embraces, because of a general quietness, for men and all other [of] their adversaries are then at sleep and rest. After their copulation in the waters, there appeareth a thick jelly, out of which the young one is found. But the land frogs are ingendered out of eggs, of whom we discourse at this present; and therefore they both suffer copulation, lay their eggs, and bring forth young ones on the land. When the egg breaketh or is hatched, there cometh forth a little black thing like a piece of flesh, which the Latins call *gyrini,* from the Greek word *gyrinos,* having no visible part of a living creature upon them, besides their eyes and their tails, and within [a] short space after their feet are formed, and their tail divided into two parts, which tail becometh their hinder-legs: wherefore when the Egyptians would describe a man that cannot move himself, and afterwards recovereth his motion, they decipher him by a frog, having his hinder-legs. The heads of these young *gyrini,* which we call in English horse-nails; because they resemble a horse-nail in their similitude, whose head is great, and the other part small, for with his tail he swimmeth. After May they grow to have feet, and if before that time they be taken out of the water they die, when they begin to have four feet. . . .

In the next place I must also show how they are likewise ingendered out of the dust of the earth by warm, æstive, and summer showers, whose life is short, and there is no use of them. . . .

And of these frogs it is that Pliny was to be understood, when he saith, that frogs in the winter time are resolved into slime, and in the summer they recover their life and substance again. It is certain also, that sometime it raineth frogs, as may appear by Philarchus and Lembus, for Lembus writeth thus: Once about Dardania, and Paeonia, it rained frogs in such plentiful measure, or rather prodigious manner, that all the houses and high ways were filled with them, and the inhabitants did first of all kill them, but afterwards perceiving no benefit thereby, they shut their doors against them, and stopped up all their lights to exclude them out of their houses, leaving no passage open so much as a frog might creep into, and yet notwithstanding all this diligence, their meat seething on the fire, or set on the table, could not be free from them, but continually they found frogs in it, so as at last they were enforced to forsake that country. It was likewise reported, that certain Indians and people of Arabia, were enforced to forsake their countries through the multitude of frogs.

Cardan seemeth to find a reason in nature for this raining of frogs, the which for the better satisfaction of the reader, I will here express as followeth: *Fiunt haec omnia ventorum ira,* and so forward in his 16.Book *De Subtilitate,* that is to say; these prodigious rains of frogs and mice, little fishes and stones, and such like things is not to be wondered at: for it cometh to pass by the rage of the winds in the tops of the mountains, or the uppermost part of the seas, which many times taketh up the dust of the earth and congealeth them into stones in the air, which afterwards fall down in rain; so also doth it take up frogs and fishes, who being above in the air, must needs fall down again. Sometimes also it taketh up the eggs of frogs and fishes, which being kept aloft in the air among the whirl-winds, and storms of showers, do there engender and bring forth young ones, which afterwards fall down upon the earth, there being no pool for them in the air. These and such like reasons are approved among the learned for natural causes of the prodigious raining of frogs.

But we read in Holy Scripture among the plagues of Egypt that frogs were sent by God to annoy them; and therefore whatsoever is the material cause, it is most certain that the wrath of God and his Almighty hand, is the making or efficient cause. . . .

The wisdom or disposition of the Egyptian frogs is much commended, for they save themselves from their enemies with singular dexterity. If they fall at any time upon a water-snake, which they know is their mortal enemy, they take in their mouths a round reed, which with an invincible strength they hold fast, never letting go, although the snake have gotten her into her mouth, for by this means the snake cannot swallow her, and so she is preserved alive. . . .

When frogs do croak above their usual custom, either more often, or more shrill than they were wont to do: they do foreshow rain and tempestuous weather.

Wherefore Tully saith in his first Book of Divination, who is it that can suspect, or once think that the little frog should know thus much, but there is in them an admirable understanding nature, constant and open to it self, but more secret and obscure to the knowledge of men, and therefore speaking to the frogs, he citeth these verses:

> And you O water-birds which dwell in streams so sweet,
>
> Do see the signs whereby the weather is foretold,
>
> Your crying voices wherewith the waters are replete,
>
> Vain sounds, absurdly moving pools and fountains cold.

And thus much for the natural use of frogs. Now followeth the magical. It is said that if a man take the tongue of a water-frog, and lay it upon the head of one that is asleep, he shall speak in his sleep, and reveal the secrets of his heart: but if he will know the secrets of a woman, then must he cut it out of the frog alive, and turn the frog away again, making certain characters upon the frog's tongue, and so lay the same upon the panting of a woman's heart, and let him ask her what questions he will, she shall answer unto him all the truth, and reveal all the secret faults that ever she hath committed. Now if this magical foolery were true, we had more need of frogs than of Justices of Peace, or Magistrates in the Common-wealth. . . .

To conclude the story of frogs, we are now to make description and narration of the toad, which is the most noble kind of frog, most venomous and remarkable for courage and strength. . . .

This toad is in all outward parts like unto a frog, the fore-feet being short, and the hinder-feet long, but the body more heavy and swelling, the colour of a blackish colour, the skin rough, viscous, and very hard, so as it is not easy to be broken with the blow of a staff. It hath many deformed spots upon it, especially black on the sides, the belly exceeding all other parts of the body, standing out in such manner, that being smitten with a staff, it yieldeth a sound as it were from a vault or hollow place. The head is broad and thick, and the colour thereof on the neather part about the neck is white, that is, somewhat pale, the back plain without bunches, and it is said, that there is a little bone growing in their sides, that hath a virtue to drive away dogs from him that beareth it about him, and is therefore called *apocynon*. The whole aspect of this toad is ugly and unpleasant. Some authors affirm that it carryeth the heart in the neck, and therefore it cannot easily be killed, except the throat thereof be cut in the middle. Their liver is very vitious, and causeth the whole body to be of ill temperament: and some say they have two livers. Their *milt* is very small, and as for their copulation and eggs, they differ nothing from frogs. . . .

A toad is of a most cold temperament, and bad constitution of nature, and it useth one certain herb wherewithal it preserveth the sight, and also resisteth the poison of spiders, whereof I have heard this credible history related, from the mouth of a true honourable man, and one of the most charitable peers of England, namely, the good Earl of Bedford, and I was requested to set it down for truth, for it may be justified by many now alive that saw the same.

It fortuned as the said Earl travailed in Bedfordshire, near unto a market-town called Owbourn, some of his company espied a toad fighting with a spider, under a hedge in a bottom, by the high-wayside, whereat they stood still, until the Earl their Lord and Master came also to behold the same; and there he saw how the spider still kept her standing, and the toad divers times went back from the spider, and did eat a piece of an herb, which to his judgment was like a plantain. At the last, the Earl having seen the toad do it often, and still return to the combat against the spider, he commanded one of his men to go, and with his dagger to cut off that herb, which he performed and brought it away. Presently after the toad returned to seek it, and not finding it according to her expectation, swelled and broke in pieces: for having received poison from the spider in the combat, nature taught her the virtue of that herb, to expel and drive it out, but wanting the herb, the poison did instantly work and destroy her. And this (as I am informed) was oftentimes related by the Earl of Bedford himself upon sundry occasions, and therefore I am the bolder to insert it into this story. . . .

"Of the Salamander"

[T]he greatest matter in the salamander to be enquired after, is whether it can live and be nourished by and in the fire, or whether it can pass through the fire without any harm, or quench and put out the same. Which opinions in the very relation and first hearing, do cross one another, for how can that either be nourished or live in the fire, which quencheth the same being put into it? Aristotle, that never saw a salamander himself, but wrote thereof by hear-say, hath given some colour to this opinion, because he writeth, *Nonulla corpora esse animalium quae igne non absumantur salamandra documento est: quae (Ut aiunt) ignem inambulans eum extinguit.* That is to say, the salamander is an evidence, that the bodies of some creatures are not wasted or consumed in the fire, for (as some say) it walketh in the fire and extinguisheth the same.

Now whether this beseemed so great a philosopher, to write upon hearsay, who took upon him to gather all natural learning into his own graunge or store-house, and out of the same to furnish both the present and all future ages, I leave it to the consideration of every indifferent reader that shall peruse this story. I for mine own part, rather judge it to be lightness in him, to insert a matter of this consequence in the discourse of this beast, without either authors or experience gathered by himself. This one thing I marvel at, why the Egyptians, when they will express or signify the man burnt, do in their hieroglyphics paint a salamander, except either fire can burn a salamander, or else contrary to all their custom, they demonstrate one contrary by another. . . .

The poison hereof is great, and not inferior to the poison of any other serpent, for sometimes by creeping upon apple-trees, it infecteth and poisoneth all the fruit, so that those which eat the same, die and languish they know not whereof: and if the heel of a man do but touch any small part or portion of the spittle of a salamander, it maketh all the hair of the body to fall off. . . .

Out of the salamander it self arise also some medicines, for it hath a septic power to eat and corrode or take away hairs, and the powder thereof cureth corns and hardness in the feet. The hear[t] tied to the wrist in a black skin, taketh away a quartain ague; and also Kiranides writeth, that being bound unto a woman's thigh, it stayeth her monthly flowers, and keepeth her barren: but this is worthily reproved for untruth, and therefore I will not commend it to the reader. And thus much for the salamander.

Thomas Browne

From *Pseudodoxia Epidemica* (1646)
"Of Frogs, Toads, and the Toadstone"

As for the stone commonly called a toad-stone, which is presumed to be found in the head of that animal, we first conceive it not a thing impossible: nor is there any substantial reason why in a toad there may not be found such hard and lapideous concretions. For the like we daily observe in the heads of fishes, as cods, carps, and perches: the like also in snails, a soft and exosseous animal, whereof in the naked and greater sort, as though she would requite the defect of a shell on their back, Nature near the head hath placed a flat white stone, or rather testaceous concretion. Which though Aldrovandus affirms, that after dissection of many, he found but in some few: yet of the great gray snails, I have not met with any that wanted it: and the same indeed so palpable, that without dissection it is discoverable by the hand.

Again, though it be not impossible, yet it is surely very rare: as we are induced to believe from some enquiry of our own, from the trial of many who have been deceived, and the frustrated search of Porta, who upon the exploreument of many, could scarce find one. Nor is it only of rarity, but may be doubted whether it be of existencie, or really any such stone in the head of a toad at all. For although lapidaries and questuary enquirers affirm it, yet the writers of minerals and natural speculators, are of another belief: conceiving the stones which bear this name, to be a mineral concretion; not to be found in animals, but in fields. And therefore Bœtius refers it to *asteria* or some kind of *lapis stellaris,* and plainly concludeth, *reperiuntur in agris, quos tamen alii in annosis ac qui diu in Arundinetis inter rubos sentesque delituerunt butfonis capitibus generari pertinaciter affirmant.**

Lastly, if any such thing there be, yet must it not, for ought I see, be taken as we receive it, for a loose and moveable stone, but rather a concretion or induration of the crany itself; for being of an earthy temper, living in the earth, and as some say feeding thereon, such indurations may sometimes happen. Thus when Brassavolus after a long search had discovered one, he affirms it was rather the forehead bone petrified, than a stone within the crany; and of this belief was Gesner. Which is also much confirmed from what is delivered in Aldrovandus, upon experiment of very many toads, whose cranies or sculs in time grew hard, and almost of a stony substance. All which considered, we must with circumspection receive those stones which commonly bear this name, much less believe the traditions, that in envy to mankind they are cast out, or swallowed down by the toad; which cannot consist with anatomy, and with the rest, enforced this censure from Bœtius, *Ab eo tempore pro nugis habui quod de Bufonio lapide, ejusque origine traditur.***

What therefore best reconcileth these divided determinations, may be a middle opinion; that of these stones some may be mineral, and to be found in

---

* "they are found in fields, though some continue to assert that they are generated from the toads' heads over time (?) and that they lay hidden in reed beds among brakes and brambles for a long time."
**"Since then I have considered as nonsense what is reported about the toad stone and its origin."

the earth; some animal, to be met with in toads, at least by the induration of their cranies. The first are many and manifold, to be found in Germany and other parts; the last are fewer in number, and in substance not unlike the stones in crabs' heads. This is agreeable unto the determination of Aldrovandus, and is also the judgment of learned Spingelius in his epistle unto Pignorius.

But these toadstones, at least very many thereof, which are esteemed among us, are at last found to be taken not out of toads' heads, but out of a fish's mouth, being handsomely contrived out of the teeth of the *lupus marinus,* a fish often taken in our Northern seas, as was publicly declared by an eminent and learned physician. But because men are unwilling to conceive so low of their toadstones which they so highly value, they may make some trial thereof by a candentorned hot iron applied unto the hollow and unpolished part thereof, whereupon if they be true stones they will not be apt to burn or afford a burnt odour, which they may be apt to do, if contrived out of animal parts or the teeth of fishes.

Concerning the generation of frogs, we shall briefly deliver that account which observation hath taught us. By frogs I understand not such as arising from putrefaction, are bred without copulation, and because they subsist not long, are called *temporarioe;* nor do I mean the little frog of an excellent parrot green, that usually sits on trees and bushes, and is therefore called *ranunculus viridis,* or *arboreus,* but hereby I understand the aquatile or water-frog, whereof in ditches and standing plashes we may behold many millions every spring in England. Now these do not as Pliny conceiveth, exclude [i.e., produce] black pieces of flesh, which after become frogs; but they let fall their spawn in the water, of excellent use in physick, and scarce unknown unto any. In this spawn of a lentous and transparent body, are to be discerned many specks, or little conglobulations, which in a small time become of deep black, a substance more compacted and terrestrious than the other; for it riseth not in distillation, and affords a powder when the white and aqueous part is exhaled. Now of this black or dusky substance is the frog at last formed; as we have beheld, including the spawn with water in a glass, and exposing it unto the sun. For that black and round substance, in a few days began to dilate and grow longer, after a while the head, the eyes, the tail to be discernable, and at last to become that which the Ancients called *gyrinus,* we a *porwigle* or tadpole. This in some weeks after becomes a perfect frog, the legs growing out before, and the tail wearing away, to supply the other behind; as may be observed in some which have newly forsaken the water; for in such, some part of the tail will be seen, but curtailed and short, not long and finny as before. A part provided them a while to swim and move in the water, that is, until such time as Nature excluded legs, whereby they might be provided not only to swim in the water, but move upon the land, according to the amphibious and mixt intention of

Nature, that is, to live in both. So that whoever observeth the first progression of the seed before motion, or shall take notice of the strange indistinction of parts in the tadpole, even when it moveth about, and how successively the inward parts do seem to discover themselves, until their last perfection; may easily discern the high curiosity of Nature in these inferiour animals, and what a long line is run to make a frog.

And because many affirm, and some deliver, that in regard it hath lungs and breatheth, a frog may be easily drowned; though the reason be probable, I find not the experiment answerable; for fastening one about a span under water, it lived almost six days. Nor is it only hard to destroy one in water, but difficult also at land: for it will live long after the lungs and heart be out; how long it will live in the seed, or whether the spawn of this year being preserved, will not arise into frogs in the next, might also be enquired: and we are prepared to trie. (Book 3, chap. 13)

"Of the Salamander"

That a salamander is able to live in flames, to endure and put out fire, is an assertion, not only of great antiquity, but confirmed by frequent, and not contemptible testimony. The Egyptians have drawn it into their hieroglyphicks, Aristotle seemeth to embrace it; more plainly Nicander, Sarenus Sammonicus, Ælian and Pliny, who assigns the cause of this effect: An animal (saith he) so cold that it extinguisheth the fire like ice. All which notwithstanding, there is on the negative, authority and experience; Sextius a physician, as Pliny delivereth, denied this effect; Dioscorides affirmed it a point of folly to believe it; Galen that it endureth the fire a while, but in continuance is consumed therein. For experimental conviction, Mathiolus affirmeth, he saw a salamander burnt in a very short time; and of the like assertion is Amatus Lusitanus; and most plainly Pierius, whose words in his hieroglyphicks are these: "Whereas it is commonly said that a salamander extinguisheth fire, we have found by experience, that it is so far from quenching hot coals, that it dieth immediately therein." As for the contrary assertion of Aristotle, it is but by hear say, as common opinion believeth, *Hæc enim (ut aiunt) ignem ingrediens, eum extinguit;** and therefore there was no absurdity in Galen, when as a septical medicine he commended the ashes of a salamander; and magicians in vain from the power of this tradition, at the burning of towns or houses expect a relief from salamanders.

---

*the belief is that she (i.e., a salamander), meeting with fire, extinguishes it.

The ground of this opinion, might be some sensible resistance of fire observed in the salamander: which being, as Galen determineth, cold in the fourth, and moist in the third degree, and having also a mucous humidity above and under the skin, by virtue thereof it may a while endure the flame: which being consumed, it can resist no more. Such an humidity there is observed in newts, or water-lizards, especially if their skins be perforated or pricked. Thus will frogs and snails endure the flame: thus will whites of eggs, vitreous or glassie flegm extinguish a coal: thus are unguents made which protect a while from the fire: and thus beside the Hirpini there are later stories of men that have passed untouched through the fire. And therefore some truth we allow in the tradition: truth according unto Galen, that it may for a time resist a flame, or as Scaliger avers, extinguish or put out a coal: for thus much will many humid bodies perform: but that it perseveres and lives in that destructive element, is a fallacious enlargement. Nor do we reasonably conclude, because for a time it endureth fire, it subdueth and extinguisheth the same, because by a cold and aluminous moisture, it is able a while to resist it: from a peculiarity of Nature it subsisteth and liveth in it. . . . (Book 3, chap. 14)

"Of the Spider and the Toad"

The antipathy between a toad and a spider, and that they poisonously destroy each other, is very famous, and solemn stories have been written of their combats; wherein most commonly the victory is given unto the spider. Of what toads and spiders it is to be understood would be considered. For the phalangium and deadly spiders, are different from those we generally behold in England. However the verity hereof, as also of many others, we cannot but desire; for hereby we might be surely provided of proper antidotes in cases which require them; but what we have observed herein, we cannot in reason conceal; who having in a glass included a toad with several spiders, we beheld the spiders without resistance to sit upon his head and pass over all his body; which at last upon advantage he swallowed down, and that in few hours, unto the number of seven. And in the like manner will toads also serve bees, and are accounted enemies unto their hives. . . . (Book 3, chap. 27)

Jan Swammerdam

From *The Book of Nature; or, The History of Insects* (1669)
"A Particular Treatise on the Generation of Frogs"

I would, on first setting out, inform the reader, that there is a much greater number of miracles, and natural secrets in the frog, than any one hath ever before thought of or discovered; as I shall evidently demonstrate, when I shall have opportunities to describe particularly the whole history of that animal; and I am now here to explain a great part of it. . . .

It is very remarkable, that living worms are almost always found in the lungs of frogs; I have often met with six in one frog. . . . These are like bristles, with sharp heads and tails, and they coil themselves up. . . . On opening these worms, immediately taken from the frog, I found in them a number of small particles wrapped up in an oblong membrane. On observing that these particles were not only all of the same size, but were likewise regularly placed as in an oviduct, my curiosity led me to examine them with a microscope, by which I at length convinced myself that this prodigious number of particles was no other than so many roundish, or oblong eggs, in which also there was a very discernable motion. This induced me to open some of them. But how great was my surprise on finding, that every egg contained a minute but perfect worm lying folded up in it, like a young serpent in its egg! Even these little worms, when extracted from their eggs, moved themselves exactly in the same manner with the parent-insect. This was an undeniable proof, that this worm was both oviparous and viviparous; thus propagating its species in the most surprising manner. But I return to my proper task; for this little animal, tho' no longer than a young frog ten days old, contained too many miracles to be published in a few pages. . . .

The eggs . . . at the time that the frog is to discharge them, break loose from the lobules of the ovary, to the insides to which they adhered, and are forced by I know not what motion thro' the cavity of the abdomen into the open mouths of the tubes; but I cannot exactly determine how much time may be requisite for this passage, as the frog is very far from being transparent like some other animals. It may not, perhaps, be impossible to observe this process in some other species: nevertheless, the knowledge of this truth could then be only analogically obtained.

About the same time we find that the testicles and spermatic vessels in the male frog fill with sperm. These animals become then so eagerly intent on the business of propagation, that they take no care in the manner of their own safety; so that they may be easily caught with the hand. I believe that they eat nothing, or very little, during this fit of lust, which lasts more or less time, according to the heat of the weather.

To carry on the intercourse of the sexes, which this great work requires, the male frog leaps upon the female, and when seated on her back, he fastens himself to her very firmly. For this reason, the Dutch country boors, with great propriety, tho' in their vulgar way, call this manner of copulation, the riding season of the frogs, as the male is carried about, riding, as it were, by the female. . . .

This wonderful copulation is performed in the following manner. As soon as the male has leapt upon the female, he throws his forelegs round her breast, encloses them so firmly, that I found it impossible to loosen them with my naked hands without wounding them; so that I found myself under a necessity of introducing an iron spatula between the female's breast and the male's legs, the better to separate them. The male most beautifully joins his toes between one another, in the same manner as people do their fingers at prayers. His head rests on that of the female, but in the hinder part, his body hangs a great deal lower than hers, as he lies so much more backwards than she does: this posture has its use, which I shall hereafter relate. The thumbs of the male's forefeet press with their thickest sides against the breast of the female, and the extreme joints of the thumbs are bent a little.

Let the female shake herself ever so much at this time, the male never lets go of his hold, even tho' she should get out of the water, so that one may carry them anywhere in this posture; which is likewise the case of garden snails, when engaged in the same business. Such is the male's eagerness to act his part, that he is not to be parted from his mate, even by pulling him forcibly from her by his hind legs. Thus these little animals swim, creep, and live together for many days successively, till the female has shed her eggs, which, at length, she does in a manner instantaneously. I observed, that when they breathed, during this intercourse, the external skin, which immediately covers the drum of the ear that lies under it, near the eyes, continually heaved up, and then fell again against that organ of hearing; and this alternate elevation and depression affords a pretty spectacle, when they both breathe, and open and shut their nostrils by turns. . . .

Thus, at last, the eggs are discharged at the female's fundament in a long stream; and the male, who has no penis, immediately fecundifies, fertilizes, or impregnates them, by an effusion of his sperm, which he likewise discharges at the anus. But as the eggs, rendered very clammy and glutinous by the white that invested them, have grown together, had been compressed in the uterus, they immediately, on being cast into the water, expand themselves into their former round form. Hence appears the necessity of the hinder part of the male's body hanging more backwards than the female's. As soon as these eggs have escaped from the female's body, between hers and the male's hinder legs, and have been impregnated by the male's sperm, the two frogs abandon each other. The male swims off, and works his fore feet as before, though they had continued so many days successively, without the least motion, in the most violent state of contraction. . . .

I own that I have endeavoured to offer something on this important subject worth the publick's acceptance; but yet I am sensible, that all this time I have been, as it were, representing with a coal the sun's meridian rays: so that this my little essay can pretend to no merit, on any other account, but that of its conformity to nature, which I hope I shall, in time, be allowed not to have misrepresented. And that time will be, when happier geniuses shall have made all these things clear and evident; for this may certainly be attained by laying aside all little thoughts of our own glory, in investigating the works of Nature, and thinking of His only, without whose assistance we could not even know any thing of them. At that happy period, the desire of writing for the sake of being talked of, will no longer prevail: we shall not then be anticipating our own praises, since all our intentions being directed to the honor of the Creator, we shall of course resist the corrupt motions of our hearts, apt to be delighted with flattery, and fond of obtaining the title of learned and ingenious men: all which I only consider as vanity of vanities, since truth is the only thing upon which we ought to depend, as on a firm foundation, and for which we ought value ourselves. Who is it amongst us, that shall discover the truth, considering our blindness in judging even of the visible objects that surround us? Hence therefore, to conclude this essay, I shall observe, that every true and valuable discovery is the gift of the Divine Grace, which God distributes as he pleases, and makes manifest at his own time. . . .

JOHN RAY

From *The Wisdom of God Manifested in the Works of the Creation* (1691)

The raining of frogs and their generation in the clouds, though it be attested by many and great authors, I look upon as utterly false and ridiculous. It seems to me no more likely, that frogs should be engendered in the clouds, than

Spanish gennets begotten by the wind; for that hath good authors too. And he that can swallow the raining of frogs, hath made a fair step towards believing, that it may rain calves also; for we read, that one fell out of the clouds in Avicen's time. Nor do they much help the matter, who say, that those frogs that appear sometimes in great multitudes after a shower, are not indeed engendered in the clouds, but coagulated of a certain sort of dust, commixt and fermented with rain-water; to which hypothesis Fromondus adheres.

But let us a little consider the generation of frogs in a natural way. 1. There are two different sexes, which must concur to their generation. 2. There is in both a great apparatus of spermatick vessels, wherein the nobler and more spirituous part of the blood, is by many digestions, concoctions, reflections, and circulations, exalted into that generous liquor we call sperm; and likewise for the preparing of the eggs. 3. There must be a copulation of the sexes, which I rather mention, because it is the most remarkable in this, that ever I observed in any animal. For they continue in *complexu venereo**at least a month indefinitely; the male all that while resting on the back of the female, clipping and embracing her with his legs about the neck and body, and holding her so fast, that if you take him out of the water, he will rather bear her whole weight, than let her go. This I observed in a couple, kept on purpose in a vessel of water, by my learned and worthy friend Mr. John Nid, Fellow of Trinity College, long since deceased. After this, the spawn must be cast into water, where the eggs lie in the midst of a copious gelly, which serves them for their first nourishment for a considerable while. And at last, the result of all is not a perfect frog, but a tadpole without any feet, and having a long tail to swim withall; in which form it continues a long time, till the limbs be grown out, and the tail fallen away, before it arrives at the perfection of a frog.

Now, if frogs can be generated spontaneously in the clouds out of vapour, or upon the earth out of dust and rain-water, what needs all this ado? To what purpose is there such an apparatus of vessels for the elaboration of the sperm and eggs; such a tedious process of generation and nutrition? This is but an idle pomp. The sun (for he is supposed to be the equivocal generant or efficient by these philosophers) could have dispatched the business in a trice: Give him but a little vapour, or a little dry dust and rain-water, he will produce you a quick frog, nay, a whole army of them, perfectly formed, and fit for all the functions of life in three minutes, nay, in the hundredth part of one minute, else must some of those frogs that were generated in the clouds fall down half-formed and imperfect, which I never heard they did; and the process of generation have been observed in the production of frogs out of dust and rain-water, which no man ever pretended to mark or discern. But that there can be no

---

*love embrace

frogs generated in the clouds, may farther be made appear, 1. From the extreme cold of the middle region of the air, where the vapours are turned into clouds, which is not at all propitious to generation. For did not so great men as Aristotle and Erasmus report it, I could hardly be induced to believe that there could be one species of insects generated in snow. 2. Because, if there were any animals engendered in the clouds, they must needs be maimed and dashed in pieces by the fall, at least such as fell in the high-ways, and upon the roofs of houses; whereas we read not of any such broken or imperfect frogs found any where. This last argument was sufficient to drive off the learned Fromondus from the belief of their generation in the clouds; but the matter of fact he takes for granted, I mean the spontaneous generation of frogs out of dust and rain-water, from an observation or experiment of his own at the gates of Tournay in Flanders, to the sight of which spectacle, he called his friends who were there present, that they might admire it with him.

> A sudden shower (saith he) falling upon the very dry dust, there suddenly appeared such an army of little frogs, leaping about every where upon the dry land, that there was almost nothing else to be seen. They were also of one magnitude and colour; neither did it appear out of what lurking places [latibula] so many myriads could creep out, and suddenly discover themselves upon the dry and dusty soil, which they hate.

But saving the reverence due to so great a man, I doubt not but they did all creep out of their holes and coverts, invited by the agreeable vapour of the rain-water. This, however unlikely it may seem, is a thousand times more probable than their instantaneous and undiscernible generation out of a little dry dust and rain-water, which also cannot have any time to mix and ferment together, which is the hypothesis he adheres to. Nay, I affirm, that it is not at all improbable; for he that shall walk out in summer-nights, when it begins to grow dark, may observe such a multitude of great toads and frogs crawling about in the high-ways, paths and avenues to houses, yards and walks of gardens and orchards, that he will wonder whence they came, or where they lurked all the winter, and all the day-time, for that then it's a rare thing to find one.

To which add, that in such frogs as we are speaking of, Monsieur Perault hath, upon dissection, often found the stomach full of meat, and the intestines of excrement; whence he justly concludes, "[t]hat they were not then first formed, but only appeared of a sudden; which is no great wonder, since upon a shower, after a drought, earth-worms and land-snails innumerable come out of their lurking places in like manner."

In confirmation of what I have here written against the spontaneous generation of frogs, either in the clouds out of vapour, or on the earth out of dust and rain-water commixed; endeavouring to prove, by force of argument, that there is no such thing; I have lately received from my learned and ingenious friend Mr. William Derham, rector of Upminster, near Rumford in Essex, a relation parallel to that of Fromondus, concerning the sudden appearance of a vast number of frogs after a shower or two of rain marching cross a sandy way, that before the rain was very dusty; and giving an account, where in all likelihood they were generated by animal parents of their own kind, and whence they did proceed. The whole narrative I shall give the reader in his own words.

> Some years ago, as I was riding forth one afternoon in Berks, I happened upon a prodigious multitude creeping cross the way. It was a sandy soil, and the way had been full of dust, by reason of a dry season that then was. But an hour or two before a refreshing fragrant shower or two of rain had laid the dust. Whereupon, what I had heard or read of the raining of frogs immediately came to my thoughts as it easily might do, there being probably as good reason then for me, as I believe any ever had before, to conclude, that these came from the clouds, or were instantaneously generated. But being prepossessed with the contrary opinion, viz. that there was no equivocal generation, I was very curious in enquiring whence this vast colony might probably come: And upon searching, I found two or three acres of land covered with this black regiment, and that they all marched the same way towards some woods, ditches, and such like cool places in their front, and from large ponds in their rear. I traced them backwards, even to the very side of one of the ponds. These ponds in spawning time always used to abound much with frogs, whose croaking I have heard at a considerable distance; and a great deal of spawn I have found there.
>
> From these circumstances I concluded, that this vast colony was bred in those ponds from whenceward they steered their course: That after their incubation (if I may so call it) or hatching by the sun, and their having passed their tadpole-state; they had lived (till that time of their migration) in the waters, or rather on the shore, among the flags, rushes, and long grass: But now being invited out by the refreshing showers, then newly fallen, which made the earth cool and moist for their march, that they left their old latibula, where perhaps they had devoured all their proper food, and were now in pursuit of food, or a more convenient habitation.

This I think not only reasonable to be concluded, but withall so easy to have been discovered by any inquisitive observer, who in former times met with the like appearance, that I cannot but admire that such sagacious philosophers, as Aristotle, Pliny, and many others since, should ever imagine frogs to fall from the clouds, or be any way instantaneously, or spontaneously generated; especially considering how openly they act their coition, produce spawn, this spawn tadpoles and tadpoles frogs.

Neither in frogs only, but also in many other creatures, as lice, flesh-flies, silk-worms, and other papilios, a uniform regular generation was very obvious, which is an argument to me of a strange pre-possession of fancy in the ages since Aristotle, not to say of carelessness and sloth. . . .

In like manner, doubtless Fromondus, had he made a diligent search, might have found out the place where those myriads of frogs, observed by him about the gates of Tournay were generated, and whence they did proceed. . . .

The discovery of the manner of the generation of these sorts of insects I earnestly recommend to all ingenious naturalists, as a matter of great moment. For if this point be but cleared, and it be demonstrated that all creatures are generated univocally by parents of their own kind, and that there is no such thing as spontaneous generation in the world, one main prop and support of atheism is taken away, and their strongest hold demolished: They cannot then exemplify their foolish hypothesis of the generation of man and other animals at first, by the like of frogs and insects at this present day.

It will be farther objected, that there have live toads been found in the midst of timber trees: nay, of stones, when they have been sawn asunder.

To this I answer, that I am not fully satisfied of the matter of fact. I am so well acquainted with the credulity of the vulgar, and the delight they, and many of the better sort too, have in telling of wonders and strange things, that I must have a thing well attested, before I can give a firm assent to it.

Since the writing hereof, the truth of these relations of live toads found in the midst of stones, hath been confirmed to me by sufficient and credible eye-witnesses, who have seen them taken out. So that there is no doubt of the matter of fact.

But yet, suppose it be true, it may be accounted for. Those animals, when young and little, finding in the stone some small hole reaching to the middle of it, might, as their nature is, creep into it, as a fit latibulum for the winter, and grow there too big to return back by the passage by which they entered, and so continue imprisoned therein for many years; a little air, by reason of the coldness of the creature, and its lying torpid there, sufficing it for respiration, and the humour of the stone, by reason it lay immoveable, and spent not, for nourishment. And I do believe, that if those who found such toads, had diligently searched, they might have discovered and traced the way whereby they entered in, or some footsteps of it. Or else there might fall down into the lapideous matter before it was concrete into a stone some small toad (or some toad spawn), which being not able to extricate itself and get out again, might remain there imprisoned till the matter about it were condensed and compacted into a stone. But however it came there, I dare confidently affirm, it was not there spontaneously generated. For else, either there was such a cavity in the stone before the toad was generated; which is altogether improbable, and *gratis dictum*, asserted without any ground; or the toad was generated in the solid stone, which is more unlikely than the other, in that the soft body of so small a creature should extend itself in such a prison, and overcome the strength and resistance of such a great and ponderous mass of solid stone.

And whereas the assertors of equivocal generation were wont to pretend the imperfection of these animals, as a ground to facilitate the belief of their spontaneous generation; I do affirm, that they are as perfect in their kind, and as much art shown in the formation of them, as of the greatest; nay more too, in the judgment of that great wit and natural historian Pliny. "*In magnis siquidem corporibus,*" (saith he) "*aut certè majoribus facilis officina sequaci materia fuit; in his tam parvis atque tam nullis, quæ ratio, quanta vis, quam inextricabilis perfectio?*" [By this Pliny meant that] in the greater bodies the forge was easy, the matter being ductile and sequacious, obedient to the hand and stroke of the artificer, apt to be drawn, formed or moulded into such shapes and machines, even by clumsy fingers: But in the formation of these, such diminutive things, such nothings, what cunning and curiosity! What force and strength was requisite, there being in them such inextricable perfection!

Part II

Reclaiming Paradise
Pioneering Nature Writers

In a perfect world, as imagined by many medieval and early modern naturalists to have existed before the Fall, human beings would exercise a responsible dominion over the rest of creation, pursuing pleasing labor in a temperate climate, in friendly but commanding consort with the beasts. The presumed loss of innocence in Eden threw a psychic wrench into the inner workings of this idyllic scene and set the human mind to wondering about its place in the scheme of things. Inquiry into the natural world was guided for centuries by the belief that following this exploratory path into the field was also a way back to God. But often coupled with the conviction that the visible creation manifested divine wisdom was an assumption that the earth could nevertheless be improved through human activity and that the human cultivation of nature was part of the plan to reclaim paradise.

It was apparently during the High Middle Ages that human transformation of nature became a topic of concern. In the thirteenth century, Albert the Great observed numerous such changes during his mendicant sojourns throughout Europe and came to believe that nature can be either improved or worsened by human activity. He noticed, for example, that cultivated grains and vegetables can be larger and milder tasting than their wild counterparts, whereas, on the other hand, improper plowing, especially on slopes, can destroy the soil. Four centuries later, when the drive for human dominion was infused with tremendous new force, the leaders of the campaign had apparently lost the medieval sense that the path toward paradise was precarious and was best tread with humility. Following the lead of early-seventeenth-century advocates of human betterment through science and technology such as Francis Bacon and René Descartes, in 1691 John Ray imagined God issuing the following address, and invitation, to humankind:

> I have provided thee with materials whereupon to exercise and employ thy art and strength; I have given thee an excel-

lent instrument, the hand, accommodated to make use of them all; I have distinguished the earth into hills and vallies, and plains, and meadows, and woods; all these parts, capable of culture and improvement by thy industry.[1]

Accepting this invitation seemed to these thinkers the surest route for humankind to regain its proper place at the pinnacle of nature's economy.

By the middle of the eighteenth century, the hand of man had worked its wonders on the European landscape as never before, and during the following hundred years, the changes were even more dramatic and far-reaching. The use of simple agricultural tools was increasingly exchanged for employment with complex machinery, and the Western world began its transformation into a modern industrial society. Faced with the novel rhythms of factory work and city life, many people of this era apparently, and understandably, considered events such as the seasonal mating of frogs to be of less than pressing importance. Consequently, Gilbert White's *Natural History of Selborne,* which deals with this and many other humble natural events, was ignored for decades following its publication in 1789. As British and American society rumbled inexorably toward the middle of the nineteenth century, however, interest in the book was definitely on the rise. Although the Selborne parson's sketches of nature have utilitarian features, readers began to find in White's collection of letters to fellow naturalists a sympathetic portrait of a vanishing world, one in which nature's providential economy was of more interest than the market economy. It proved irresistible to the many searching souls who beat a path to Selborne. American writer James Russell Lowell visited in 1850 and described White's book as "the journal of Adam in Paradise."

The desire to reclaim such an arcadian paradise was a crucial stimulus to modern nature writing, which purposed to offer relief to those who had grown weary of the industrial paradise born of the Baconian vision. John Burroughs made the trip to Selborne in 1882 and remarked that to appreciate fully White's low-key letters, it is probably necessary that "the disturbing elements of the great hurly-burly outside world" not cloud the reader's consciousness, as they apparently did not weigh heavily in the writer's mind or village.[2] Not so in Henry David Thoreau's Concord. Thoreau, who consulted White's book regularly, watched the outside world penetrate his village on the tracks of the Fitchburg Railroad in 1850 and read the effects of growing commercialism in the "quiet desperation" of many of his contemporaries. His writings, along with those of Burroughs and William Henry Hudson, offer critiques, both direct and indirect, of the modern inner and outer landscape, with its characteristic alienation of human beings from the natural world.

An aspect of modernity that each of these pioneering nature writers adopted, even though not embracing it wholeheartedly, was modern science. Carolus Linnaeus praised John Bartram as "the greatest natural botanist in the world," and both John Bartram and his son William, whose natural history expertise grew to surpass even that of his father, benefited greatly from Linnaeus's innovative classification scheme. In 1852, Thoreau made a note to himself in his journal to "read Linnaeus at once," which he did, and made good use of the Swede's work in his careful perusal of nature. Similarly, in the later nineteenth century, Burroughs and Hudson found Charles Darwin's vision liberating and enriching inasmuch as it reinserted human life into the natural order.

But each of these naturalists saw the bare science of the day as ultimately deficient for the task of regaining a glimpse of paradise. In the service of that goal, they transfused into their science the warmth of human sentiment and into nature the integrated vitality of a living organism. William Bartram recognized a mechanical aspect in animal life but spoke also of a "more essential principle, which secretly operates within," and produced a seminal travel account that made a particularly strong impression on Romantic kindred spirits in England such as Samuel Taylor Coleridge and William Wordsworth. Thoreau preferred the methods of the older naturalists, such as Edward Topsell, to those of most modern scientific inquirers because the former "sympathize with the creatures they describe." And he argued that "surely the most important part of an animal is its *anima,* its vital spirit, on which is based its character and all the peculiarities by which it most concerns us," but which most scientific texts omitted, treating an animal as if it were simply inert material.[3] Burroughs, who called himself a "nature-lover" rather than a "scientific-naturalist," criticized the coldness of purely scientific knowledge by asking the question: "What is knowledge without enjoyment, without love? It is sympathy, appreciation, emotional experience, which refine and elevate and breathe into exact knowledge the breath of life."[4]

GILBERT WHITE

From *The Natural History of Selborne* (1789)

Advertisement

The author of the following Letters takes the liberty, with all proper deference, of laying before the public his idea of *parochial history,* which, he thinks, ought

Figure 4
European common frog *(Rana temporaria).*
(From *The Riverside Natural History,* 1888, p. 342.)

to consist of natural productions and occurrences as well as antiquities. He is also of opinion that if stationary men would pay some attention to the districts on which they reside, and would publish their thoughts respecting the objects that surround them, from such materials might be drawn the most complete county-histories, which are still wanting in several parts of this kingdom, and in particular in the county of Southampton. . . .

If the writer should at all appear to have induced any of his readers to pay a more ready attention to the wonders of the Creation, too frequently overlooked as common occurrences; or if he should by any means, through his researches, have lent a helping hand towards the enlargement of the boundaries of historical and topographical knowledge; or if he should have thrown some small light upon ancient customs and manners and especially on those that were monastic, his purpose will be fully answered. But if he should not have been successful in any of these his intentions, yet there remains this consolation behind—that these his pursuits, by keeping the body and mind employed, have, under Providence, contributed to much health and cheerfulness of spirits, even to old age: and, what still adds to his happiness, have led him to the knowledge of a circle of gentlemen whose intelligent communications, as they have afforded him much pleasing information, so, could he flatter himself with a continuation of them, would they ever be deemed a matter of singular satisfaction and improvement.

Letter XVII to Selborne, June 18, 1768
Thomas Pennant, Esquire

Dear Sir,

On Wednesday last arrived your agreeable letter of June the 10th. It gives me great satisfaction to find that you pursue these studies still with such vigour, and are in such forwardness with regard to reptiles and fishes.

The reptiles, few as they are, I am not acquainted with, so well as I could wish, with regard to their natural history. There is a degree of dubiousness and obscurity attending the propagation of this class of animals, sometimes analogous to that of the *cryptogamia* in the sexual system of plants: and the case is the same as regards some of the fishes: as the eel, etc.

The method in which toads procreate and bring forth seems to me very much in the dark. Some authors say that they are viviparous: and yet Ray classes them among his oviparous animals; and is silent with regard to the manner of their bringing forth. Perhaps they may be ἔσω μεν ωοτόκοι ἔξω δε ζωοτόκοι,* as is known to be the case with the viper.

The copulation of frogs (or at least the appearance of it; for Swammerdam proves that the male has no *penis intrans*) is notorious to everybody: because we see them sticking upon each other's backs for a month together in spring: and yet I never saw, or read, of toads being observed in the same situation. It is strange that the matter with regard to the venom of toads has not yet been settled. That they are not noxious to some animals is plain: for ducks, buzzards, owls, stone curlews, and snakes, eat them, to my knowledge, with impunity. And I well remember the time, but was not eye-witness to the fact (though numbers of persons were), when a quack, at this village, ate a toad to make the country people stare; afterwards he drank oil.

I have been informed also, from undoubted authority, that some ladies (ladies you will say of peculiar taste) took a fancy to a toad, which they nourished summer after summer, for many years, till he grew to a monstrous size, with the maggots which turn to flesh flies. The reptile used to come forth every evening from an hole under the garden-steps; and was taken up, after supper, on the table to be fed. But at last a tame raven, kenning him as he put forth his head, gave him such a severe stroke with his horny beak as put out one eye. After this accident the creature languished for some time and died.

I need not remind a gentleman of your extensive reading of the excellent account there is from Mr. Derham, in Ray's *Wisdom of God in the Creation* (p. 365), concerning the migration of frogs from their breeding ponds. In this

---

* internally (?) they are oviparous but externally (?) they are viviparous

account he at once subverts that foolish opinion of their dropping from the clouds in rain; showing that it is from the grateful coolness and moisture of those showers that they are tempted to set out on their travels, which they defer till those fall. Frogs are as yet in their tadpole state; but in a few weeks, our lanes, paths, fields, will swarm for a few days with myriads of these emigrants, no larger than my little finger nail. Swammerdam gives a most accurate account of the method and situation in which the male impregnates the spawn of the female. How wonderful is the œconomy of Providence with regard to the limbs of so vile a reptile! While it is aquatic it has a fish-like tail, and no legs: as soon as the legs sprout, the tail drops off as useless, and the animal betakes itself to the land.

Merret, I trust, is widely mistaken when he advances that the *rana arborea* is an English reptile; it abounds in Germany and Switzerland.

It is to be remembered that the *salamandra aquatica* of Ray (the water-newt or eft) will frequently bite at the angler's bait, and is often caught on his hook. I used to take it for granted that the *salamandra aquatica* was hatched, lived, and died in the water. But John Ellis, Esq., F.R.S. (the coralline Ellis) asserts, in a letter to the Royal Society, dated June 5th, 1766, in his account of the *mud inguana,* an amphibious *bipes,* from South Carolina, that the water-eft, or newt, is only the larva of the land-eft, as tadpoles are of frogs. Lest I should be suspected to misunderstand his meaning, I shall give it in his own words. Speaking of the *opercula* or covering to the gills of the *mud inguana,* he proceeds to say that "[t]he forms of these pennated coverings approach very near to what I have some time ago observed in the larva or aquatic state of our English *lacerta,* known by the name of eft, or newt; which serve them for coverings to their gills, and for fins to swim with while in this state; and which they lose, as well as the fins of their tails, when they change their state, and become land animals, as I have observed, by keeping them alive for some time myself."

Linnaeus, in his *Systema Naturæ,* hints at what Mr. Ellis advances more than once. . . .

Letter XVIII to Thomas Pennant, Esquire   Selborne, July 27, 1768

Dear Sir,

I received your obliging and communicative letter of June the 28th, while I was on a visit at a gentleman's house, where I had neither books to turn to, nor leisure to sit down, to return you an answer to many queries, which I wanted to resolve in the best manner that I am able. . . .

In my visit I was not very far from Hungerford, and did not forget to make some inquiries concerning the wonderful method of curing cancers by means of toads. Several intelligent persons, both gentry and clergy, do, I find, give a great deal of credit to what was asserted in the papers: and I myself dined with a clergyman who seemed to be persuaded that what is related is matter of fact;

but, when I came to attend to his account, I thought I discerned circumstances which did not a little invalidate the woman's story of the manner in which she came by her skill. She says of herself "that, labouring under a virulent cancer, she went to some church where there was a vast crowd: on going into a pew, she was accosted by a strange clergyman; who, after expressing compassion for her situation, told her that if she would make such an application of living toads as is mentioned she would be well." Now is it likely that this unknown gentleman should express so much tenderness for this single sufferer, and not feel any for the many thousands that daily languish under this terrible disorder? Would he not have made use of this invaluable nostrum for his own emolument; or, at least, by some means of publication or other, have found a method of making it public for the good of mankind? In short, this woman (as it appears to me) having set up for a cancer-doctor, finds it expedient to amuse the country with this dark and mysterious relation.

The water-eft has not, that I can discern, the least appearance of any gills; for want of which it is continually rising to the surface of the water to take in fresh air. I opened a big-bellied one indeed, and found it full of spawn. Not that this circumstance at all invalidates the assertion that they are larvae: for the larvae of insects are full of eggs, which they excude the instant they enter their last state. The water-eft is continually climbing over the brims of the vessel, within which we keep it in water, and wandering away: and people every summer see numbers crawling out of the pools where they are hatched, up the dry banks. There are varieties of them, differing in colour; and some have fins up their tail and back, and some have not. . . .

William Bartram

From *Travels Through North and South Carolina, Georgia, East and West Florida* (1791)

Introduction

The attention of a traveller should be particularly turned, in the first place, to the various works of Nature, to mark the distinctions of the climates he may explore, and to offer such useful observations on the different productions as may occur. Men and manners undoubtedly hold the first rank—whatever may contribute to our existence is also of equal importance, whether it be found in the animal or vegetable kingdom; neither are the various articles, which tend to promote the happiness and convenience of mankind, to be disregarded. How far the writer of the following sheets has succeeded in furnishing information on these subjects, the reader will be capable of determining. From the advantages the journalist enjoyed under his father John Bartram, botanist to the king of Great Britain, and fellow of the Royal Society, it is hoped that his labours will present new as well as useful information to the botanist and zoologist.

This world, as a glorious apartment of the boundless palace of the sovereign Creator, is furnished with an infinite variety of animated scenes, inexpressibly beautiful and pleasing, equally free to the inspection and enjoyment of all his creatures.

Perhaps there is not any part of creation, within the reach of our observations, which exhibits a more glorious display of the Almighty hand, than the vegetable world: such a variety of pleasing scenes, ever changing throughout the seasons, arising from various causes, and assigned each to the purpose and use determined. . . .

The animal creation also excites our admiration, and equally manifests the almighty power, wisdom, and beneficence of the Supreme Creator and Sovereign Lord of the universe; some in their vast size and strength, as the mammoth, the elephant, the whale, the lion, and alligator; others in agility; others in their beauty and elegance of colour, plumage, and rapidity of flight, having the faculty of moving and living in the air; others for their immediate and indispensable use and convenience to man, in furnishing means for our clothing and sustenance, and administering to our help in the toils and labours of life: how wonderful is the mechanism of these finely formed self-moving beings, how complicated their system, yet what unerring uniformity prevails through every tribe and particular species! the effect we see and contemplate, the cause is invisible, incomprehensible; how can it be otherwise? when we cannot see the end or origin of a nerve or vein, while the divisibility of matter or fluid, is infinite. We admire the mechanism of a watch, and the fabric of a piece of brocade, as being the production of art; these merit our admiration, and must excite our esteem for the ingenious artist or modifier; but nature is the work of God omnipotent; and an elephant, nay even this world, is comparatively but a very minute part of his works. If then the visible, the mechanical part of the animal creation, the mere material part, is so admirably beautiful, harmonious, and incomprehensible, what must be the intellectual system? that inexpressibly more essential principle, which secretly operates within? that which animates the inimitable machines, which gives them motion, impowers them to act, speak, and perform, this must be divine and immortal?

I am sensible that the general opinion of philosophers has distinguished the moral system of the brute creature from that of mankind, by an epithet which implies a mere mechanical impulse, which leads and impels them to necessary actions, without any premeditated design or contrivance; this we term instinct, which faculty we suppose to be inferior to reason in man.

The parental and filial affections seem to be as ardent, their sensibility and attachment as active and faithful, as those observed in human nature.

When travelling on the east coast of the isthmus of Florida, ascending the south Musquito river, in a canoe, we observed numbers of deer and bears, near the banks, and on the islands of the river: the bears were feeding on the fruit of the dwarf creeping Chamærops; (this fruit is of the form and size of dates, and is delicious and nourishing food) we saw eleven bears in the course of the day, they seemed no way surprised or affrighted at the sight of us. In the evening, my hunter, who was an excellent marksman, said that he would shoot one of them for the sake of the skin and oil, for we had plenty and variety of provisions in our bark. We accordingly, on sight of two of them, planned our approaches as artfully as possible, by crossing over to the opposite shore, in order to get under cover of a small island; this we cautiously coasted round, to a point, which we apprehended would take us within shot of the bears; but here finding ourselves at too great a distance from them, and discovering that we must openly show ourselves, we had no other alternative to effect our purpose, but making oblique approaches. We gained gradually on our prey by this artifice, without their noticing us: finding ourselves near enough, the hunter fired, and laid the target dead on the spot where she stood; when presently the other, not seeming the least moved at the report of our piece, approached the dead body, smelled, and pawed it, and appearing in agony, fell to weeping and looking upwards, then towards us, and cried out like a child. Whilst our boat approached very near, the hunter was loading his rifle in order to shoot the survivor, which was a young cub, and the slain supposed to be the dam. The continual cries of this afflicted child, bereft of its parent, affected me very sensibly; I was moved with compassion, and charging myself as if accessary to what now appeared to be a cruel murder, endeavoured to prevail on the hunter to save its life, but to no effect! for by habit he had become insensible to compassion towards the brute creation: being now within a few yards of the harmless devoted victim, he fired, and laid it dead upon the body of the dam.

If we bestow but very little attention to the economy of the animal creation, we shall find manifest examples of premeditation, perseverance, resolution, and consummate artifice, in order to effect their purposes. . . .

Part II: Chapter 10

. . . Since I have begun to mention the animals of these regions, this may be a proper place to enumerate the other tribes which I observed during my perigrinations. I shall begin with the frogs (Ranae).

(1) The largest frog known in Florida and on the sea coast of Carolina, is about eight or nine inches in length from the nose to the extremity of the toes: they are of a dusky brown or black colour on the upper side, and their belly or under side white, spotted and clouded with dusky spots of various size and figure; their legs and thighs also are variegated with transverse ringlets, of dark brown or black; and they are yellow and green about their mouth and lips. They live in wet swamps and marshes, on the shores of large rivers and lakes; their voice is loud and hideous, greatly resembling the grunting of a swine; but not near as loud as the voice of the bull frog of Virginia and [Pennsylvania]: neither do they arrive to half their size, the bull frog being frequently eighteen inches in length, and their roaring as loud as that of a bull.

(2) The bell frog, so called because their voice is fancied to be exactly like the sound of a loud cow bell. This tribe being very numerous, and uttering their voices in companies or by large districts, when one begins another answers; thus the sound is caught and repeated from one to another, to a great distance round about, causing a surprising noise for a few minutes, rising and sinking according as the wind sits, when it nearly dies away, or is softly kept up by distant districts or communities: thus the noise is repeated continually, and as one becomes familiarised to it, is not unmusical, though at first, to strangers, it seems clamorous and disgusting.

(3) A beautiful green frog inhabits the grassy, marshy shores of these large rivers. They are very numerous, and their noise exactly resembles the barking of little dogs, or the yelping of puppies: these likewise make a great clamour, but as their notes are fine, and uttered in chorus, by separate bands or communities, far and near, rising and falling with the gentle breezes, affords a pleasing kind of music.

(4) There is besides this a less green frog, which is very common about houses: their notes are remarkably like that of young chickens: these raise their chorus immediately preceding a shower of rain, with which they seem delighted.

(5) A little grey speckled frog is in prodigious numbers in and about the ponds and savannas on high land, particularly in Pine forests: their language or noise is also uttered in chorus, by large communities or separate bands; each particular note resembles the noise made by striking two pebbles together under the surface of the water, which when thousands near you utter their notes at the same time, and is wafted to your ears by a sudden flow of wind, is very surprising, and does not ill resemble the rushing noise made by a vast quantity of gravel and pebbles together, at once precipitated from a great height.

(6) There is yet an extreme diminutive species of frogs, which inhabits the grassy verges of ponds in savannas: these are called savanna crickets, are of a dark ash or dusky colour, and have a very picked nose. At the times of very great rains, in the autumn, when the savannas are in a manner inundated, they

From *Travels Through North and South Carolina, Georgia, East and West Florida* (1791) 49

Figure 5
Southern cricket frog *(Acris gryllus)*. (From *The Riverside Natural History,* 1888, p. 337.)

are to be seen in incredible multitudes clambering up the tall grass, weed, &c. round the verges of the savannas, bordering on the higher ground; and by an inattentive person might be taken for spiders or other insects. Their note is very feeble, not unlike the chattering of young birds or crickets.

(7) The shad frog, so called in [Pennsylvania] from their appearing and croaking in the spring season, at the time the people fish for shad: this is a beautiful spotted frog, of a slender form, five or six inches in length from the nose to the extremities; of a dark olive green, blotched with clouds and ringlets of a dusky colour: these are remarkable jumpers and enterprising hunters, leaving their ponds to a great distance in search of prey. They abound in rivers, swamps and marshes, in the Southern regions; in the evening and sultry summer days, particularly in times of drought, are very noisy; and at some distance one would be almost persuaded that there were assemblies of men in serious debate. These have also a sucking or clucking noise, like that which is made by sucking in the tongue under the roof of the mouth. These are the kinds of water frogs that have come under my observation; yet I am persuaded that there are yet remaining several other species.

(8) The high land frogs, commonly called toads, are of two species, the red and black. The former, which is of a reddish brown or brick colour, is the largest, and may weigh upwards of one pound when full grown: they have a disagreeable look, and when irritated, they swell and raise themselves up on their four legs and croak, but are no ways venomous or hurtful to man. The other species are one third less, and of a black or dark dusky colour. The legs and thighs of both are marked with blotches and ringlets of a darker colour, which appear more conspicuous when provoked: the

smaller black species are the most numerous. Early in the spring season, they assemble by numberless multitudes in the drains and ponds, when their universal croaking and shouts are great indeed, yet in some degree not unharmonious. After this breeding time they crawl out of the water and spread themselves all over the country. Their spawn being hatched in the warm water, the larva is there nourished, passing through the like metamorphoses as the water frogs; and as soon as they obtain four feet, whilst yet no larger than crickets, they leave the fluid nursery-bed, and hop over the dry land after their parents.

The food of these amphibious creatures, when out of the water, is every kind of insect, reptile, &c. they can take, even ants and spiders; nature having furnished them with an extreme long tongue, which exudes a viscid or glutinous liquid, they being secreted under covert, spring suddenly upon their prey, or dart forth their tongue as quick as lightning, and instantly drag into their devouring jaws the unwary insect. But whether they prey upon one another, as the water frogs do, I know not. . . .

Henry David Thoreau
From the *Journal* (1857–1860)

Sept. 12, 1857. Saturday. P.M.—To Owl Swamp (Farmer's) In an open part of the swamp, started a very large wood frog, which gave one leap and squatted still. I put down my finger, and, though it shrank a little at first, it permitted me to stroke it as long as I pleased. Having passed, it occurred to me to return and cultivate its acquaintance. To my surprise, it allowed me to slide my hand under it and lift it up, while it squatted cold and moist on the middle of my palm, panting naturally. I brought it close to my eye and examined it. It was very beautiful seen thus nearly, not the dull dead-leaf color which I had imagined, but its back was like burnished bronze armor defined by a varied line on each side, where, as it seemed, the plates of armor united. It had four or five dusky bars which matched exactly when the legs were folded, showing that the painter applied his brush to the animal when in that position, and reddish-orange soles to its delicate feet. There was a conspicuous dark-brown patch along the side of the head, whose upper edge passed directly through the eye horizontally, just above its centre, so that the pupil and all below were dark and the upper portion of the iris golden. I have since taken up another in the same way. . . .

Nov. 2, 1857 ... Are not the wood frogs philosophers who walk (?) in these groves? Methinks I imbibe a cool, composed, frog-like philosophy when I behold them....

Apr. 18, 1858 ... Frogs are strange creatures. One would describe them as peculiarly wary and timid, another as equally bold and imperturbable. All that is required in studying them is patience. You will sometimes walk a long way along a ditch and hear twenty or more leap in one after another before you, and see where they rippled the water, without getting sight of one of them. Sometimes, as this afternoon the two *R. fontinalis,* when you approach a pool or spring a frog hops in and buries itself at the bottom. You sit down on the brink and wait patiently for his reappearance. After a quarter of an hour or more he is sure to rise to the surface and put out his nose quietly without making a ripple, eying you steadily. At length he becomes as curious about you as you can be about him. He suddenly hops straight toward [you], pausing within a foot, and takes a near and leisurely view of you. Perchance you may now scratch its nose with your finger and examine it to your heart's content, for it is become as imperturbable as it was shy before. You conquer them by superior patience and immovableness; not by quickness, but by slowness; not by heat, but by coldness....

May 1, 1858 ... A warm and pleasant day, reminding me of the 3d of April when the *R. halecina* waked up so suddenly and generally, and now, as then, apparently a new, allied frog is almost equally wide awake,—the one of last evening (and before).

When I am behind Cheney's this warm and still afternoon, I hear a voice calling to oxen three quarters of a mile distant, and I know it to be Elijah Wood's. It is wonderful how far the *individual* proclaims himself. Out of the thousand millions of human beings on this globe, I know that this sound was made by the lungs and larynx and lips of E. Wood, am as sure of it as if he nudged me with his elbow and shouted in my ear. He can impress himself on the very atmosphere, then, can launch himself a mile on the wind, through trees and rustling sedge and over rippling water, associating with a myriad sounds, and yet arrive distinct at my ear; and yet this creature that is felt so far, that was so noticeable, lives but a short time, quietly dies and makes no more noise that I know of. I can tell him, too, with my eyes by the very gait and motion of him half a mile distant. Far more wonderful his purely spiritual influence,—that after the lapse of thousands of years you may still detect the individual in the turn of a sentence or the tone of a thought!! E. Wood has a peculiar way of modulating the air, imparts to it peculiar vibrations, which several times when standing near him I have noticed, and now a vibration, spreading far and wide over the fields and up and down the river, reaches me and maybe hundreds of others, which we all know to have been produced by Mr.

Wood's pipes. However, E. Wood is not a match for a little peeping hylodes in this respect, and there is no peculiar divinity in this.

The inhabitants of the river are peculiarly wide awake this warm day,— fishes, frogs, and toads, from time to time,—and quite often I hear a tremendous rush of a pickerel after his prey. They are peculiarly active, maybe after the *Rana palustris,* now breeding. It is a perfect frog and toad day. I hear the stertorous notes of last evening from all sides of the river at intervals, but most from the grassiest and warmest or most sheltered and sunniest shores. I get sight of ten or twelve *Rana palustris* and catch three of them. One apparent male utters one fine, sharp squeak when caught. Also see by the shore one apparent young bullfrog (?), with bright or vivid light green just along its jaws, a dark line between this and jaws, and a white throat; head, brown above. This is the case with one I have in the firkin, which I think was at first a dull green. These are the only kinds I find sitting along the river. The *Rana palustris* is the prevailing one, and I suppose it makes the *halecina*-like sound described last night. They will be silent for a long time. You will see perhaps one or two snouts and eyes above the surface, then at last may hear a coarsely purring croak, often rapid and as if it began with a *p,* at a distance sounding softer and like *tut tut, tut tut, tut,* lasting a second or two; and then, perchance, others far and near will be excited to utter similar sounds, and all the shore seems alive with them. However, I do not as yet succeed to see one make this sound. Then there may be another pause of fifteen or thirty minutes.

The *Rana palustris* leaves a peculiar strong scent on the hand, which reminds me of days when I went a-fishing for pickerel and used a frog's leg for bait. When I try to think what it smells like, I am inclined to say that it might be the bark of some plant. It is disagreeable. Some are in the water, others on the shore.

I do not see a single *R. halecina.* What has become of the thousands with which the meadows swarmed a month ago? They have given place to the *R. palustris.* Only their spawn, mostly hatched and dissolving, remains, and I expect to detect the spawn of the *palustris* soon.

I find many apparent young bullfrogs in the shaded pools on the Island Neck. There is one good-sized bullfrog among them.

The toads are so numerous, some sitting on all sides, that their ring is a continuous sound throughout the day and night, if it is warm enough, as it now is, except perhaps in the morning. It is as uninterrupted to the ear as the rippling breeze or the circulations of air itself, for when it dies away on one side it swells again on another, and if it should suddenly cease men would exclaim at the pause, though they might not have noticed the sound itself. . . .

MAY 6, 1858 . . . About 9 P.M. I went to the edge of the river to hear the frogs. It was a warm and moist, rather foggy evening, and the air full of the ring of the toad, the peep of the hylodes, and the low *growling* croak or stertoration of the *Rana palustris.* Just there, however, I did not hear much of the toad, but rather from the road, but I heard the steady peeping of innumerable hylodes for a background to the *palustris* snoring, further over the meadow. There was a universal snoring of the *R. palustris* all up and down the river on each side, the very sounds that mine made in my chamber last night, and probably it began in earnest last evening on the river. It is a hard, dry, unmusical, *fine* watchman's-rattle-like stertoration, swelling to a speedy conclusion, lasting say some four or five seconds usually. The rhythm of it is like that of the toads' ring, but not the sound. This is considerably like that of the tree toad, when you think of it critically, after all, but is not so musical or sonorous as that even. There is an occasional more articulate, querulous, or rather quivering, alarm note such as I have described (May 2d). Each shore of the river now for its whole length is all alive with this stertorous purring. It is such a sound as I make in my throat when I imitate the growling of wild animals. I have heard a little of it at intervals for a week, in the warmest days, but now at night it [is] universal all along the river. If the note of the *R. halecina,* April 3d, was the first awakening of the river meadows, this is the second,—considering the hylodes and toads less (?) peculiarly of the river meadows. Yet how few distinguished this sound at all, and I know not one who can tell what frog makes it, though it is almost as universal as the breeze itself. The sounds of those three reptiles now fill the air, especially at night. The toads are most regardless of the light, and regard less a cold day than the *R. palustris* does. In the mornings now, I hear no *R. palustris* and no hylodes, but a few toads still, but now, at night, all ring together, the toads ringing through the day, the hylodes beginning in earnest toward night and the *palustris* at evening. I think that the different epochs in the revolution of the seasons may perhaps be best marked by the notes of reptiles. They express, as it were, the very feelings of the earth or nature. They are perfect thermometers, hygrometers, and barometers. . . .

MARCH 24, 1859. P.M.—DOWN RAILROAD Southeast wind. Begins to sprinkle while I am sitting in Laurel Glen, listening to hear the earliest wood frogs croaking. I think they get under weigh a little earlier, i.e., you will hear many of them sooner than you will hear many hylodes. Now, when the leaves get to be dry and rustle under your feet, dried by the March winds, the peculiar dry note, *wurrk wurrk wur-r-r-k wurk* of the wood frog is heard faintly by ears on the alert, borne up from some unseen pool in a woodland hollow which is open to the influences of the sun. It is a singular sound for awakening Nature to make, associated with the first warmer days, when you sit in some sheltered place in the woods amid the dried leaves. How moderate on her first awakening, how little demonstrative! You may sit half an hour before you will hear another. You doubt if the season will be long enough for such Oriental and luxurious slowness. But they get on, nevertheless, and by tomorrow, or in a day or two, they croak louder and more frequently. Can you ever be sure that you have heard the very first wood frog in the township croak?

Ah! how weather-wise must he be! There is no guessing at the weather with him. He makes the weather in his degree; he encourages it to be mild. The weather, what is it but the temperament of the earth? and he is wholly of the earth, sensitive as its skin in which he lives and of which he is a part. His life relaxes with the thawing ground. He pitches and tunes his voice to chord with the rustling leaves which the March wind has dried. Long before the frost is quite out, he feels the influence of the spring rains and the warmer days. His is the very voice of the weather. He rises and falls like quicksilver in the thermometer. You do not perceive the spring so surely in the actions of men, their lives are so artificial. They may make more fire or less in their parlors, and their feelings accordingly are not good thermometers. The frog far away in the wood, that burns no coal nor wood, perceives more surely the general and universal changes. . . .

FEB. 17, 1860. P.M. . . . We cannot spare the very lively and lifelike descriptions of some of the old naturalists. They sympathize with the creatures which they describe. Edward Topsell in his translation of Conrad Gesner, in 1607, called "The History of Four-footed Beasts," says of the antelopes that "they are bred in India and Syria, near the river Euphrates," and then— which enables you to realize the living creature and its habitat—he adds, "and delight much to drink of the cold water thereof." The beasts which most modern naturalists describe do not *delight* in anything, and their water is neither hot nor cold. Reading the above makes you want to go and drink of the Euphrates yourself, if it is warm weather. I do not know how much of his spirit he owes to Gesner, but he proceeds in his translation to say that "they have horns growing forth of the crown of their head, which are very long and sharp; so that Alexander affirmed they pierced through the shields of his soldiers, and fought with them very irefully: at which time his company slew as he travelled to India, eight thousand five hundred and fifty, which great slaughter may be the occasion why they are so rare and seldom seen to this day."

Now here *something* is described at any rate; it is a real account, whether of a real animal or not. . . .

Though some beasts are described in this book which have no existence as I can learn but in the imagination of the writers, they really have an existence there, which is saying not a little, for most of our modern authors have not imagined the actual beasts which they presume to describe. . . .

FEB. 18, 1860 . . . Gesner (unless we owe it to the translator) has a livelier conception of an animal which has no existence, or of an action which was never performed, than most naturalists have of what passes before their eyes. The ability to report a thing *as if [it] had occurred,* whether it did or not, is surely important to a describer. They do not half tell a thing because you might expect them to but half believe it. I feel, of course, very ignorant in a museum. I know nothing about the things which they have there,—no more than I should know my friends in the tomb. I walk amid those jars of bloated creatures which they label frogs, a total stranger, without the least froggy thought being suggested. Not one of them can croak. They leave behind all life they that enter there, both frogs and men. . . .

If you have undertaken to write the biography of an animal, you will have to present to us the living creature, *i.e.,* a result which no man can understand, but only in his degree report the impression made on him.

Science in many departments of natural history does not pretend to go beyond the shell; *i.e.,* it does not get to animated nature at all. A history of animated nature must itself be animated.

The ancients, one would say, with their gorgons, sphinxes, satyrs, mantichora, etc., could imagine more than existed, while the moderns cannot imagine so much as exists.

John Burroughs

From *Pepacton* (1904)
"The Tree-Toad"

We can boast a greater assortment of toads and frogs in this country than can any other land. What a chorus goes up from our ponds and marshes in spring! The like of it cannot be heard anywhere else under the sun. In Europe it would certainly have made an impression upon the literature. An attentive ear will detect first one variety, then another, each occupying the stage from three or four days to a week. The latter part of April, when the little peeping frogs are in full chorus, one comes upon places, in his drives or walks late in the day, where the air fairly palpitates with sound; from every little marshy hollow and spring run there rises an impenetrable maze or cloud of shrill musical voices. After the peepers, the next frog to appear is the clucking frog, a rather small, dark-brown frog, with a harsh, clucking note, which later in the season becomes the well-known brown wood frog. Their chorus is heard for a few days only, while their spawn is being deposited. In less than a week it ceases, and I never hear them again till the next April. As the weather gets warmer, the toads take to the water, and set up that long-drawn musical tr-r-r-r-r-r-ing note. The voice of the bullfrog, who calls, according to the boys, "jug o' rum," "jug o' rum," "pull the plug," "pull the plug," is not heard much before June. The peepers, the clucking frog, and the bullfrog are the only ones that call in chorus. The most interesting and the most shy and withdrawn of all our frogs and toads is the tree toad,—the creature that, from the old apple or cherry tree, or red cedar, announces the approach of rain, and baffles your every effort to see or discover it. It has not (as some people imagine) exactly the power of the chameleon to render itself invisible by assuming the color of the object it perches upon, but it sits very close and still, and its mottled back, of different shades of ashen gray, blends it perfectly with the bark of nearly every tree. The only change in its color I have ever noticed is that it is lighter on a light-colored tree, like the beech or soft maple, and darker on the apple, or cedar, or pine. Then it is usually hidden in some cavity or hollow of the tree, when its voice appears to come from the outside.

Most of my observations upon the habits of this creature run counter to the authorities I have been able to consult on the subject.

In the first place, the tree toad is nocturnal in its habits, like the common toad. By day it remains motionless and concealed; by night it is as alert and active as an owl, feeding and moving about from tree to tree. I have never known one to change its position by day, and never knew one to fail to do so by night. Last summer one was discovered sitting against a window upon a climbing rosebush. The house had not been occupied for some days, and when the curtain was drawn the toad was discovered and closely observed. His light gray color harmonized perfectly with the unpainted woodwork of the house. During the day he never moved a muscle, but next morning he was gone. A friend of mine caught one, and placed it under a tumbler on his table at night, leaving the edge of the glass raised about the eighth of an inch to admit the air. During the night he was awakened by a strange sound in his room. Pat, pat, pat went some object, now here, now there, among the furniture, or upon the walls and doors. On investigating the matter, he found that by some means his tree toad had escaped from under the glass, and was leaping in a very lively manner about the room, producing the sound he had heard when it alighted upon the door, or wall, or other perpendicular surface.

The home of the tree toad, I am convinced, is usually a hollow limb or other cavity in the tree; here he makes his headquarters, and passes most of the day. For two years a pair of them frequented an old apple-tree near my house, occasionally sitting at the mouth of a cavity that led into a large branch, but usually their voices were heard from within the cavity itself. On one occasion, while walking in the woods in early May, I heard the voice of a tree toad but a few yards from me. Cautiously following up the sound, I decided, after some delay, that it proceeded from the trunk of a small soft maple; the tree was hollow, the entrance to the interior being a few feet from the ground. I could not discover the toad, but was so convinced that it was concealed in the tree, that

I stopped up the hole, determined to return with an axe, when I had time, and cut the trunk open. A week elapsed before I again went to the woods, when, on cutting into the cavity of the tree, I found a pair of tree toads, male and female, and a large, shelless snail. Whether the presence of the snail was accidental, or whether these creatures associated together for some purpose, I do not know. The male toad was easily distinguished from the female by its large head, and more thin, slender, and angular body. The female was much the more beautiful, both in form and color. The cavity, which was long and irregular, was evidently their home; it had been nicely cleaned out, and was a snug, safe apartment.

The finding of the two sexes together, under such circumstances and at that time of the year, suggests the inquiry whether they do not breed away from the water, as others of our toads are known at times to do, and thus skip the tadpole state. I have several times seen the ground, after a June shower, swarming with minute toads, out to wet their jackets. Some of them were no larger than crickets. They were a long distance from the water, and had evidently been hatched on the land, and had never been polliwogs. Whether the tree toad breeds in trees or on the land, yet remains to be determined. Another fact in the natural history of this creature, not set down in the books, is that they pass the winter in a torpid state in the ground, or in stumps and hollow trees, instead of in the mud of ponds and marshes, like true frogs, as we have been taught. The pair in the old apple-tree above referred to, I heard on a warm, moist day late in November, and again early in April. On the latter occasion, I reached my hand down into the cavity of the tree and took out one of the toads. It was the first I had heard, and I am convinced it had passed the winter in the moist, mud-like mass of rotten wood that partially filled the cavity. It had a fresh, delicate tint, as if it had not before seen the light that spring. The president of a Western college writes in "Science News" that two of his students found one in the winter in an old stump which they demolished; and a person whose veracity I have no reason to doubt sends me a specimen that he dug out of the ground in December while hunting for Indian relics. The place was on the top of a hill, under a pine-tree. The ground was frozen on the surface, and the toad was, of course, torpid.

During the present season, I obtained additional proof of the fact that the tree toad hibernates on dry land. The 12th of November was a warm, spring-like day; wind southwest, with slight rain in the afternoon,—just the day to bring things out of their winter retreats. As I was about to enter my door at dusk, my eye fell upon what proved to be the large tree toad in question, sitting on some low stone-work at the foot of a terrace a few feet from the house. I paused to observe his movements. Presently he started on his travels across the yard toward the lawn in front. He leaped about three feet at a time,

Figure 6
Tree frogs *(Hyla palmate)*. (From *The Riverside Natural History*, 1888, p. 338.)

with long pauses between each leap. For fear of losing him as it grew darker, I captured him, and kept him under the coal sieve till morning. He was very active at night trying to escape. In the morning, I amused myself with him for some time in the kitchen. I found he could adhere to a window-pane, but could not ascend it; gradually his hold yielded, till he sprang off on the casing. I observed that, in sitting upon the floor or upon the ground, he avoided bringing his toes in contact with the surface, as if they were too tender or delicate for such coarse uses, but sat upon the hind part of his feet. Said toes had a very bungling, awkward appearance at such times; they looked like hands encased in gray woolen gloves much too large for them. Their round, flattened ends, especially when not in use, had a comically helpless look.

After a while I let my prisoner escape into the open air. The weather had grown much colder, and there was a hint of coming frost. The toad took the hint at once, and, after hopping a few yards from the door to the edge of a grassy bank, began to prepare for winter. It was a curious proceeding. He went into the ground backward, elbowing himself through the turf with the sharp joints of his hind legs, and going down in a spiral manner. His progress was very slow: at night I could still see him by lifting the grass; and as the weather changed again to warm, with southerly winds before morning, he stopped digging entirely. The next day I took him out, and put him into a bottomless tub sunk into the ground and filled with soft earth, leaves, and leaf mould, where he passed the winter safely, and came out fresh and bright in the spring.

The little peeping frogs lead a sort of arboreal life, too, a part of the season, but they are quite different from the true tree toads above described. They appear to leave the marshes in May, and to take to the woods or bushes. I have never seen them on trees, but upon low shrubs. They do not seem to be climbers, but perchers. I caught one in May, in some low bushes a few rods from the swamp. It perched upon the small twigs like a bird, and would leap about among them, sure of its hold every time. I was first attracted by its piping. I brought it home, and it piped for one twilight in a bush in my yard and then was gone. I do not think they pipe much after leaving the water. I have found them early in April upon the ground in the woods, and again late in the fall.

In November, 1879, the warm, moist weather brought them out in numbers. They were hopping about everywhere upon the fallen leaves. Within a small space I captured six. Some of them were the hue of the tan-colored leaves, probably Pickering's *hyla,* and some were darker, according to the locality. Of course they do not go to the marshes to winter, else they would not wait so late in the season. I examined the ponds and marshes, and found bullfrogs buried in the mud, but no peepers.

W. H. HUDSON

From *The Book of a Naturalist* (1919)
Chapter 8: "The Toad as Traveller"

One summer day I sat myself down on the rail of a small wooden footbridge—a very old bridge it looked, bleached to a pale grey colour with grey, green, and yellow lichen growing on it, and very creaky with age, but the rail was still strong enough to support my weight. The bridge was at the hedgeside, and the stream under it flowed out of a thick wood over the road and into a marshy meadow on the other side, overgrown with coarse tussocky grass. It was a relief to be in that open sunny spot, with the sight of water and green grass and blue sky before me, after prowling for hours in the wood—a remnant of the old Silchester forest—worried by wood-flies in the dense undergrowth. These same wood-flies and some screaming jays were all the wild creatures I had seen, and I would now perhaps see something better at that spot.

It was very still, and for some time I saw nothing, until my wandering vision lighted on a toad travelling towards the water. He was right out in the middle of the road, a most dangerous place for him, and also difficult to travel in, seeing that it had a rough surface full of loosened stones, and was very dusty. His progress was very slow; he did not hop, but crawled laboriously for about five inches, then sat up and rested four or five minutes, then crawled and rested again. When I first caught sight of him he was about forty yards from the water, and looking at him through my binocular when he sat up and rested I could see the pulsing movements of his throat as though he panted with fatigue, and the yellow eyes on the summit of his head gazing at that delicious coolness where he wished to be. If toads can see things forty yards away the stream was visible to him, as he was on that part of the road which sloped down to the stream.

Lucky for you, old toad, thought I, that it is not market day at Basingstoke or somewhere with farmers and small general dealers flying about the country in their traps, or you would be flattened by a hoof or a wheel long before the end of your pilgrimage.

By and by another creature appeared and caused me to forget the toad. A young water-vole came up stream, swimming briskly from the swampy meadow on the other side of the road. As he approached I tapped the wood with my stick to make him turn back, but this only made him swim faster towards me, and determined to have my own way I jumped down and tried to stop him, but he dived past the stick and got away where he wanted to be in the wood, and I resumed my seat.

There was the toad, when I looked his way, just about where I had last seen him, within perhaps a few inches. Then a turtledove flew down, alighting within a yard of the water, and after eyeing me suspiciously for a few moments advanced and took one long drink and flew away. A few minutes later I heard a faint complaining and whining sound in or close to the hedge on my left hand, and turning my eyes in that direction caught sight of a stoat, his head and neck visible, peeping at me out of the wood; he was intending to cross the road, and seeing me sitting there hesitated to do so. Still having come that far he would not turn back, and by and by he drew himself snake-like out of the concealing herbage, and was just about to make a dash across the road when I tapped sharply on the wood with my stick and he fled back into cover. In a few seconds he appeared again, and I played the same trick on him with the same result; this was repeated about four times, after which he plucked up courage enough to make his dash and was quickly lost in the coarse grass by the stream on the other side.

Then a curious thing happened: flop, flop, flop, went vole following vole, escaping madly from their hiding-places along the bank into the water, all swimming for dear life to the other side of the stream. Their deadly enemy did not swim after them, and in a few seconds all was peace and quiet again.

And when I looked at the road once more, the toad was still there, still travelling, painfully crawling a few inches, then sitting up and gazing with his yellow eyes over the forty yards of that weary *via dolorosa* which still had to be got over before he could bathe and make himself young for ever in that river of life. Then all at once the feared and terrific thing came upon him: a farmer's trap, drawn by a fast trotting horse, suddenly appeared at the bend of the road and came flying down the slope. That's the end of you, old toad, said I, as the horse and trap came over him; but when I had seen them cross the ford and vanish from sight at the next bend, my eyes went back, and to my amazement there sat my toad, his throat still pulsing, his prominent eyes still gazing forward. The four dread hoofs and two shining wheels had all missed him; then at long last I took pity on him, although vexed at having to play providence to a toad, and getting off the rail I went and picked him up, which made him very angry. But when I put him in the water he expanded and floated for a few moments with legs spread out, then slowly sank his body and remained with just the top of his head and the open eyes above the surface for a little while, and finally settled down into the cooler depths below.

It is strange to think that when water would appear to be so much to these water-born and amphibious creatures they yet seek it for so short a period in each year, and for the rest of the time are practically without it! The toad comes to it in the love season, and at that time one is often astonished at the number of toads seen gathered in some solitary pool, where perhaps not a toad has been seen for months past, and with no other water for miles around. The fact is, the solitary pool has drawn to itself the entire toad population of the surrounding country, which may comprise an area of several square miles. Each toad has his own home or hermitage somewhere in that area, where he spends the greater portion of the summer season practically without water excepting in wet weather, hiding by day in moist and shady places, and issuing forth in the evening. And there too he hibernates in winter. When spring returns he sets out on his annual pilgrimage of a mile or two, or even a greater distance, travelling in the slow, deliberate manner of the one described, crawling and resting until he arrives at the sacred pool—his Tipperary. They arrive singly and are in hundreds, a gathering of hermits from the desert places, drunk with excitement, and filling the place with noise and commotion. A strange sound, when at intervals the leader or precentor or bandmaster for the moment blows himself out into a wind instrument—a fairy bassoon, let us say, with a tremble to it—and no sooner does he begin than a hundred more join in; and the sound, which the scientific books describe as "croaking," floats far and wide, and produces a beautiful, mysterious effect on a still evening when the last heavy-footed labourer has trudged home to his tea, leaving the world to darkness and to me.

In England we are almost as rich in toads as in serpents, since there are two species, the common toad, universally distributed, and the rarer natterjack, abundant only in the south of Surrey. The breeding habits are the same in both species, the concert-singing included, but there is a difference in the *timbre* of their voices, the sound produced by the natterjack being more resonant and musical to most ears than that of the common toad.

The music and revels over, the toads vanish, each one taking his own road, long and hard to travel, to his own solitary home. Their homing instinct, like that of many fishes and of certain serpents that hibernate in numbers together, and of migrating birds, is practically infallible. They will not go astray, and the hungriest raptorial beasts, foxes, stoats, and cats, for example, decline to poison themselves by killing and devouring them.

In the late spring or early summer one occasionally encounters a traveller on his way back to his hermitage. I met one a mile or so from the valley of the Wylie, half-way up a high down, with his face to the summit of Salisbury Plain. He was on the bank at the side of a deep narrow path, and was resting on the velvety green turf, gay with little flowers of the chalk-hills—eye-bright, squinancy-wort, daisies, and milkwort, both white and blue.

The toad, as a rule, strikes one as rather an ugly creature, but this one sitting on the green turf, with those variously coloured fairy flowers all about him, looked almost beautiful. He was very dark, almost black, and with his shining topaz eyes had something of the appearance of a yellow-eyed black cat. I sat down by his side and picked him up, which action he appeared to regard as an unwarrantable liberty on my part; but when I placed him on my knee and began stroking his blackish corrugated back with my finger-tips his anger vanished, and one could almost imagine his golden eyes and wide lipless mouth smiling with satisfaction.

A good many flies were moving about at that spot—a pretty fly whose name I do not know, a little bigger than a house-fly, all a shining blue, with head and large eyes a bright red. These flies kept lighting on my hand, and by and by I cautiously moved a hand until a fly on it was within tongue-distance of the toad, whereupon the red tongue flicked out like lightning and the fly vanished. Again the process was repeated, and altogether I put over half-a-dozen flies in his way, and they all vanished in the same manner, so quickly that the action eluded my sight. One moment and a blue and red-headed fly was on my hand sucking the moisture from the skin, and then, lo! he was gone, while the toad still sat there motionless on my knee like a toad carved out of a piece of black stone with two yellow gems for eyes.

After helping him to a dinner, I took him off my knee with a little trouble, as he squatted close down, desiring to stay where he was, and putting him back among the small flowers to get more flies for himself if he could, I went on my way.

It is easy to establish friendly relations with these lowly creatures, amphibious and reptiles, by a few gentle strokes with the finger-tips on the back. Shortly after my adventure with this toad I was visiting a naturalist friend, who told me of an adventure he had had with a snake. He was out walking with his wife near his home among the Mendips when they spied the snake basking in the sun on the turf, and at the same moment the snake saw them and began quietly gliding away. But they succeeded in overtaking and capturing it, and, although it was a large snake and struggled violently to escape, they soon quieted it down by stroking its back with their fingers. They kept and played with it for half an hour, then put it down, whereupon it went away, but quite slowly, almost as if reluctant to leave them.

So far this was a common experience; I have tamed many grass-snakes in the same way, and the only smooth snake I have ever captured in England was made tame in about ten minutes by holding it on my knee and stroking it. In the instance related by my friend, it would appear that the tameness does not always vanish as soon as the creature finds itself free again. About three days after the incident I have related he was again walking with his wife, and they again found the snake at the same spot, whereupon he, anxious to capture it again, made a dash at it, but the snake on this occasion made no attempt to escape, and when picked up did not struggle. They again kept it some time, caressing it with their fingers, then releasing it as before; later they saw their snake on several occasions, when it acted in the same way, allowing itself to be taken up and kept as long as it was wanted, and then, when released, going very slowly away.

That one first delightful experience of having its back stroked with finger-tips had made a tame snake of it.

Part III

Telling Naturalistic Tales
Scientific Essayists

When Charles Darwin returned to England in 1836 from his five-year voyage on the HMS *Beagle,* he was a changed man. He had set out with a copy of *The Natural History of Selborne* in his bags and apparently envisioned a career as a parson-naturalist in the mold of Gilbert White, immersing himself in the edifying study of providential natural harmony. But his experiences and observations in the South Pacific, particularly on the Galápagos Islands, revealed to him a darker side of both human nature and Mother Nature than he had encountered in the misty isles. After two years back home, he remarked in a notebook on the "transmutation of species" that "[i]t is difficult to believe in the dreadful but quiet war of organic beings going on [in] the peaceful woods & smiling fields."[1] Darwin saw a grimace behind that peaceful smile and gradually convinced the scientific world to confront it face to face. Even though the Darwinian natural economy was marked by a competitiveness foreign to earlier naturalists, the encompassing community of creatures depicted by White never lost its appeal for Darwin. He visited Selborne in the 1850s, and the germ of his final book, *The Formation of Vegetable Mould Through the Action of Earthworms,* came from White's discussion of worms a century before. Both men recognized that even animals of the soil and swamps played important roles in nature. In the following excerpts from Darwin's writings, his observation of frogs and toads progresses from his acquaintance with them on the South American mainland to his awareness of their absence on oceanic islands, with a tentative hypothesis of the cause, to his later use of this evidence in the *Origin of Species* to make a more confident theoretical statement.

Darwin was quite happy to leave the defense of his grand theory to some of his more pugnacious followers, and the foremost duo was Thomas Henry Huxley and Ernst Haeckel. The latter, coiner of the term *oecologie,* added an unusual twist to evolutionary theory by placing the mechanism of natural selection within a larger organismic vision of the progress of life. He was also

Figure 7
"Batrachia." (From Ernst Haeckel, *Art Forms in Nature*, 1904.)

an influential artist, producing a famous 1874 biological illustration showing comparable embryological development in a salamander, a chicken, and a human being as well as elaborate paintings of plants and animals, such as the one on the previous page. Huxley, though a fast friend of Haeckel, would have none of the German's nature philosophy, which posited a pervasive unity of matter and spirit. The English Darwinian had abandoned his belief in a divine Author of nature when he concluded that Darwin's theory had destroyed the argument from design. And he thought it senseless to believe that one can know anything about the ultimate nature of the universe, inventing the term *agnostic* to describe his viewpoint. He saw the bodies of animals, including human bodies, as automata that emit incidental mental states in the same way that a clock rings a bell, and he took pains to prove the point in a famous 1870 lecture titled "Has a Frog a Soul?" Historian William Irvine has said that in this lecture, Huxley "crucified the unredemptive batrachian on paraffin in order once more to crucify the immortality of the soul on the hard facts of physiology."[2] Huxley subsequently fleshed out his argument and presented it again four years later in his controversial and influential essay "On the Hypothesis that Animals are Automata, and Its History."

The gradual acceptance of Darwinism, especially as championed by the likes of Huxley, had the effect of expelling any remaining spirits from the worldview of mainstream biological science, giving it a more purely naturalistic, in contrast to *super*naturalistic, flavor. This agreement to limit one's inquiries and explanations to the phenomena of the natural world as it is laid out in space and time was, of course, a proposition to which most twentieth-century scientists, in one form or another, readily subscribed. The scientific essayists included in this part communicate their perspective and passion for natural history in an engaging literary style, though with less of the personal element than in most true nature writing.

In the early twentieth century, Scottish biologist J. Arthur Thomson said in the preface to his *Biology of the Seasons* that he strove to combine "the sympathetic feeling of the old naturalists, such as Gilbert White, with Darwin's dominant sense of correlation and evolution." His "Tale of Tadpoles" from this book of season lore also invokes the popular nineteenth-century idea, propounded especially by Ernst Haeckel, that "ontogeny recapitulates phylogeny," meaning in this case that the metamorphosis of frogs "re-enacts the epoch-making colonization of the dry land." Julian Huxley, grandson of T. H. Huxley, was one of the leading architects of the modern synthesis of Darwinian theory with modern genetics. Although he viewed the world through essentially naturalistic eyes, his essay reveals his belief in the inherently creative and opportunistic nature of evolution, which he saw both in biological achievements such as amphibian and reptilian life and in the higher human world of cultural inventions such as wagons and automobiles.

American anthropologist and natural historian Loren Eiseley was one of the leading scientific essayists of the latter twentieth century. His startling story of an archetypal "dance of the frogs" wonderfully depicts the young anthropologist's discomfiture when faced with a glimpse into a realm apparently beyond the reach of science. It brings to mind a passage from Thoreau that Eiseley used to open *The Immense Journey* (1957), his first book: "Man can not afford to be a naturalist, to look at Nature directly, but only with the side of his eye. He must look through and beyond her." Stephen Jay Gould, another immensely influential writer, takes a brief excursion into the history of Darwinism as an entrée into a discussion of the fascinating gastric brooding frog of Australia, the emergence of which Gould sees as evidence of evolution heading off in creative directions, which occasionally "become seeds of major innovations and floods of diversity in life's history." In this case, however, it was not to be so: the gastric brooding frog was apparently one of the early casualties in the worldwide amphibian decline. Finally, although ecologist David Scott's recent essay on the marbled salamander focuses on an amphibian apparently not in decline, it stresses the particular value today of long-term field studies of population dynamics. It also highlights the devotion of this researcher to plumbing the mysteries of these creatures, these "magic wells," as Edward O. Wilson would say, to their seemingly inexhaustible depths.

Charles Darwin

From *The Voyage of the Beagle* (1845)
"Rio de Janeiro"

The climate, during the months of May and June, or the beginning of winter, was delightful. The mean temperature, from observations taken at nine o'clock, both morning and evening, was only 72°. It often rained heavily, but the drying southerly winds soon again rendered the walks pleasant. One morning, in the course of six hours, 1.6 inches of rain fell. As this storm passed over the forests which surround the Corcovado, the sound produced by the drops pattering on the countless multitude of leaves was very remarkable; it could be heard at the distance of a quarter of a mile, and was like the rushing of a great body of water. After the hotter days, it was delicious to sit quietly in the garden and watch the evening pass into night. Nature, in these climes, chooses her vocalists from more humble performers than in Europe. A small frog, of the genus Hyla, sits on a blade of grass about an inch above the surface of the water, and sends forth a pleasing chirp: when several are together they sing in harmony on different notes. I had some difficulty in catching a specimen of this frog. The genus Hyla has its toes terminated by small suckers; and I found this animal could crawl up a pane of glass, when placed absolutely perpendicular. Various cicada and crickets, at the same time, keep up a ceaseless shrill cry, but which, softened by the distance, is not unpleasant. Every evening after dark this great concert commenced; and often have I sat listening to it, until my attention has been drawn away by some curious passing insect. . . .

"Bahia Blanca"

Amongst the Batrachian reptiles, I found only one little toad (*Phryniscus nigricans*), which was most singular from its colour. If we imagine, first, that it had been steeped in the blackest ink, and then, when dry, allowed to crawl over a board, freshly painted with the brightest vermilion, so as to colour the soles of its feet and parts of its stomach, a good idea of its appearance will be gained. If it had been an unnamed species, surely it ought to have been called *Diabolicus,* for it is a fit toad to preach in the ear of Eve. Instead of being nocturnal in its habits, as other toads are, and living in damp obscure recesses, it crawls during the heat of the day about the dry sand-hillocks and arid plains, where not a single drop of water can be found. It must necessarily depend on the dew for its moisture; and this probably is absorbed by the skin, for it is known, that these reptiles possess great powers of cutaneous absorption. At Maldonado, I found one in a situation nearly as dry as at Bahia Blanca, and thinking to give it a great treat, carried it to a pool of water; not only was the little animal unable to swim, but, I think without help it would soon have been drowned.

"Galápagos Archipelago"

We will now turn to the order of reptiles, which gives the most striking character to the zoology of these islands. The species are not numerous, but the numbers of individuals of each species are extraordinarily great. There is one small lizard belonging to a South American genus, and two species (and probably more) of the *Amblyrhynchus*—a genus confined to the Galápagos Islands.

There is one snake which is numerous; it is identical, as I am informed by M. Bibron, with the *Psammophis Temminckii* from Chile. Of sea-turtle I believe there is more than one species; and of tortoises there are as we shall presently show, two or three species or races. Of toads and frogs there are none: I was surprised at this, considering how well suited for them the temperate and damp upper woods appeared to be. It recalled to my mind the remark made by Bory St. Vincent, namely, that none of this family are found on any of the volcanic islands in the great oceans. As far as I can ascertain from various works, this seems to hold good throughout the Pacific, and even in the large islands of the Sandwich archipelago. Mauritius offers an apparent exception, where I saw the *Rana Mascariensis* in abundance: this frog is said now to inhabit the Seychelles, Madagascar, and Bourbon; but on the other hand, Du Bois, in his voyage of 1669, states that there were no reptiles in Bourbon except tortoises; and the officer of du Roi asserts that before 1768 it had been attempted, without success, to introduce frogs into Mauritius—I presume for the purpose of eating: hence it may well be doubted whether this frog is an aboriginal of these islands. The absence of the frog family in the oceanic islands is the more remarkable, when contrasted with the case of lizards, which swarm on most of the smallest islands. May this difference not be caused, by the greater facility with which the eggs of lizards, protected by calcareous shells, might be transported through salt-water, than could the slimy spawn of frogs?

From *The Origin of Species* (1859)
Introduction

When on board H.M.S. 'Beagle,' as naturalist, I was much struck with certain facts in the distribution of the organic beings inhabiting South America, and in the geological relations of the present to the past inhabitants of that continent. These facts, as will be seen in the latter chapters of this volume, seemed to throw some light on the origin of species—that mystery of mysteries, as it has been called by one of our greatest philosophers. On my return home, it occurred to me, in 1837, that something might perhaps be made out on this question by patiently accumulating and reflecting on all sorts of facts which could possibly have any bearing on it. After five years' work I allowed myself to speculate on the subject, and drew up some short notes; these I enlarged in 1844 into a sketch of the conclusions, which then seemed to me probable: from that period to the present day I have steadily pursued the same object. I hope that I may be excused for entering on these personal details, as I give them to show that I have not been hasty in coming to a decision. . . .

This Abstract, which I now publish, must necessarily be imperfect. I cannot here give references and authorities for my several statements; and I must trust to the reader reposing some confidence in my accuracy. No doubt errors will have crept in, though I hope I have always been cautious in trusting to good authorities alone. I can here give only the general conclusions at which I have arrived, with a few facts in illustration, but which, I hope, in most cases will suffice. No one can feel more sensible than I do of the necessity of hereafter publishing in detail all the facts, with references, on which my conclusions have been grounded; and I hope in a future work to do this. For I am well aware that scarcely a single point is discussed in this volume on which facts cannot be adduced, often apparently leading to conclusions directly opposite to those at which I have arrived. A fair result can be obtained only by fully stating and balancing the facts and arguments on both sides of each question; and this is here impossible. . . .

Chapter 13: "Geographical Distribution"

With respect to the absence of whole orders of animals on oceanic islands, Bory St. Vincent long ago remarked that batrachians (frogs, toads, newts) are never found on any of the many islands with which the great oceans are studded. I have taken pains to verify this assertion, and have found it true, with the exception of New Zealand, New Caledonia, the Andaman Islands, and perhaps the Solomon Islands and the Seychelles. But I have already remarked that it is doubtful whether New Zealand and New Caledonia ought to be classed as oceanic islands; and this is still more doubtful with respect to the Andaman and Solomon groups and the Seychelles. This general absence of frogs, toads, and newts on so many true oceanic islands cannot be accounted for by their physical conditions: indeed it seems that islands are peculiarly fitted for these animals; for frogs have been introduced into Madeira, the Azores, and Mauritius, and have multiplied so as to become a nuisance. But as these animals and their spawn are immediately killed (with the exception, as far as known, of one Indian species) by sea-water, there would be great difficulty in their transportal across the sea, and therefore we can see why they do not exist on strictly oceanic islands. But why, on the theory of creation, they should not have been created there, it would be very difficult to explain. . . .

Thomas Henry Huxley

"On the Hypothesis That Animals Are Automata, and Its History" (1874)

The first half of the seventeenth century is one of the great epochs of biological science. For though suggestions and indications of the conceptions which took definite shape, at that time, are to be met with in works of earlier date, they are little more than the shadows which coming truth casts forward; men's knowledge was neither extensive enough, nor exact enough, to show them the solid body of fact which threw these shadows.

But, in the seventeenth century, the idea that the physical processes of life are capable of being explained in the same way as other physical phenomena, and, therefore, that the living body is a mechanism, was proved to be true for certain classes of vital actions; and, having thus taken firm root in irrefragable fact, this conception has not only successfully repelled every assault which has been made upon it, but has steadily grown in force and extent of application, until it is now the expressed or implied fundamental proposition of the whole doctrine of scientific Physiology. . . .

The attempt to reduce the endless complexities of animal motion and feeling to law and order is, at least, as important a part of the task of the physiologist as the elucidation of what are sometimes called the vegetative processes. Harvey did not make this attempt himself; but the influence of his work upon the man who did make it is patent and unquestionable. This man was René Descartes, who, though by many years Harvey's junior, died before him; and yet, in his short span of fifty-four years, took an undisputed place, not only among the chiefs of philosophy, but amongst the greatest and most original of mathematicians; while, in my belief, he is no less certainly entitled to the rank of a great and original physiologist; inasmuch as he did for the physiology of motion and sensation that which Harvey had done for the circulation of the blood, and opened up that road to the mechanical theory of these processes, which has been followed by all his successors. . . .

Now, if by some accident, a man's spinal cord is divided, his limbs are paralyzed, so far as his volition is concerned, below the point of injury, and he is incapable of experiencing all those states of consciousness which, in his uninjured state, would be excited by irritations of those nerves which come off below the injury. If the spinal cord is divided in the middle of the back, for example, the skin of the feet may be cut, or pinched, or burned, or wetted with vitriol, without any sensation of touch, or of pain, arising in consciousness. So far as the man is concerned, therefore, the part of the central nervous system which lies beyond the injury is cut off from consciousness. . . .

If the spinal cord of a frog is cut across, so as to provide us with a segment separated from the brain, we shall have a subject parallel to the injured man, on which experiments can be made without remorse; as we have a right to conclude that a frog's spinal cord is not likely to be conscious when a man's is not.

Now the frog behaves just as the man did. The legs are utterly paralysed, so far as voluntary movement is concerned; but they are vigorously drawn up to the body when any irritant is applied to the foot. But let us study our frog a little farther. Touch the skin of the side of the body with a little acetic acid, which gives rise to all the signs of great pain in an uninjured frog. In this case, there can be no pain, because the application is made to a part of the skin supplied with nerves which come off from the cord below the point of section; nevertheless, the frog lifts up the limb of the same side, and applies the foot to rub off the acetic acid; and, what is still more remarkable, if the limb be held so that the frog cannot use it, it will, by-and-by, move the limb of the other side, turn it across the body, and use it for the same rubbing process. It is impossible that the frog, if it were in its entirety and could reason, should perform actions more purposive than these: and yet we have most complete assurance that, in this case, the frog is not acting from purpose, has no consciousness, and is a mere insensible machine.

But now suppose that, instead of making a section of the cord in the middle of the body, it had been made in such a manner as to separate the hindermost division of the brain from the rest of the organ, and suppose the foremost two-thirds of the brain entirely taken away. The frog is then absolutely devoid of any spontaneity; it sits upright in the attitude which a frog habitually assumes; and it will not stir unless it is touched; but it differs from the frog which I have just described in this, that, if it be thrown into the water, it begins to swim, and swims just as well as the perfect frog does. But swimming requires the combination and successive coordination of a great number of muscular actions. And we are forced to conclude, that the impression made upon the sensory nerves of the skin of the frog by the contact with the water into which it is thrown, causes the transmission to the central nervous apparatus of an impulse, which sets going a certain machinery by which all the muscles of swimming are brought into play in due coordination. If the frog be stimulated by some irritating body, it jumps or walks as well as the complete frog can do. The simple sensory impression, acting through the machinery of the cord, gives rise to these complex combined movements.

It is possible to go a step farther. Suppose that only the anterior division of the brain—so much of it as lies in front of the "optic lobes"—is removed. If that operation is performed quickly and skilfully, the frog may be kept in a state of full bodily vigour for months, or it may be for years; but it will sit unmoved. It sees nothing; it hears nothing. It will starve sooner than feed itself, although food put into its mouth is swallowed. On irritation, it jumps or walks; if thrown into the water it swims. If it be put on the hand, it sits there, crouched, perfectly quiet, and would sit there for ever. If the hand be inclined very gently and slowly, so that the frog would naturally tend to slip off, the creature's fore paws are shifted on to the edge of the hand, until he can just prevent himself from falling. If the turning of the hand be slowly continued, he mounts up with great care and deliberation, putting first one leg forward and then another, until he balances himself with perfect precision upon the edge; and, if the turning of the hand is continued, over he goes through the needful set of muscular operations, until he comes to be seated in security, upon the back of the hand. The doing of all this requires a delicacy of coordination, and a precision of adjustment of the muscular apparatus of the body, which are only comparable to those of a rope-dancer. To the ordinary influences of light, the frog, deprived of its central hemispheres, appears to be blind. Nevertheless, if the animal be put upon a table, with a book at some little distance between it and the light, and the skin of the hinder part of its body is then irritated, it will jump forward, avoiding the book by passing to the right or left of it. Although the frog, therefore, appears to have no sensation of light, visible objects act through its brain upon the motor mechanism of its body.

It is obvious, that had Descartes been acquainted with these remarkable results of modern research, they would have furnished him with far more powerful arguments than he possessed in favour of his view of the automatism of brutes. The habits of a frog, leading its natural life, involve such simple adaptations to surrounding conditions, that the machinery which is competent to do so much without the intervention of consciousness, might well do all. And this argument is vastly strengthened by what has been learned in recent times of the marvellously complex operations which are performed mechanically, and to all appearance without consciousness, by men, when, in consequence of injury or disease, they are reduced to a condition more or less comparable to that of a frog, in which the anterior part of the brain has been removed. . . .

As I have endeavoured to show, we are justified in supposing that something analogous to what happens in ourselves takes place in the brutes, and that the affections of their sensory nerves give rise to molecular changes in the brain, which again give rise to, or evolve, the corresponding states of consciousness. Nor can there be any reasonable doubt that the emotions of brutes, and such ideas as they possess, are similarly dependent upon molecular brain changes. Each sensory impression leaves behind a record in the structure of the brain—an "ideagenous" molecule, so to speak, which is competent, under certain conditions, to reproduce, in a fainter condition, the state of consciousness which corresponds with that sensory impression; and it is these "ideagenous molecules" which are the physical basis of memory.

It may be assumed, then, that molecular changes in the brain are the causes of all the states of consciousness of brutes. Is there any evidence that these states of consciousness may, conversely, cause those molecular changes which give rise to muscular motion? I see no such evidence. The frog walks, hops, swims, and goes through his gymnastic performances quite as well without consciousness, and consequently without volition, as with it; and, if a frog, in his natural state, possesses anything corresponding with what we call volition, there is no reason to think that it is anything but a concomitant of the molecular changes in the brain which form part of the series involved in the production of motion. . . .

The hypothesis that brutes are conscious automata is perfectly consistent with any view that may be held respecting the often discussed and curious question whether they have souls or not; and, if they have souls, whether those souls are immortal or not. It is obviously harmonious with the most literal adherence to the text of Scripture concerning "the beast that perisheth;" but it is not inconsistent with the amiable conviction ascribed by Pope to his "untutored savage," that when he passes to the happy hunting-grounds in the sky, "his faithful dog shall bear him company." If the brutes have consciousness and no souls, then it is clear that, in them, consciousness is a direct function of material changes; while, if they possess immaterial subjects of consciousness, or souls, then, as consciousness is brought into existence only as the consequence of molecular motion of the brain, it follows that it is an indirect product of material changes. The soul stands related to the body as the bell of a clock to the works, and consciousness answers to the sound which the bell gives out when it is struck. . . .

Figure 8
Skeleton of frog *(Ceratophrys dorsata)*. (From *The Riverside Natural History,* 1888, p. 326.)

It is quite true that, to the best of my judgment, the argumentation which applies to brutes holds equally good of men; and, therefore, that all states of consciousness in us, as in them, are immediately caused by molecular changes of the brain-substance. It seems to me that in men, as in brutes, there is no proof that any state of consciousness is the cause of change in the motion of the matter of the organism. If these positions are well based, it follows that our mental conditions are simply the symbols in consciousness of the changes which take place automatically in the organism; and that, to take an extreme illustration, the feeling we call volition is not the cause of a voluntary act, but the symbol of that state of the brain which is the immediate cause of that act. We are conscious automata, endowed with free will in the only intelligible sense of that much-abused term—inasmuch as in many respects we are able to do as we like—but none the less parts of the great series of causes and effects which, in unbroken continuity, composes that which is, and has been, and shall be—the sum of existence.

J. Arthur Thomson

From *The Biology of the Seasons* (1911)
"The Tale of Tadpoles"

The frogs are among the earliest heralds of the spring, for although their croaking (in March or earlier) may not be particularly attractive in our ears, it has the same deep *motif* as the nightingale's song. It is a "love"-call. Awakening after a winter's lethargy and fasting, the frogs creep out of the mud of the pond and call to one another. They unite in couples, and the eggs laid by the female in the water are fertilised by the male just as they are laid. These eggs form the familiar masses of "frog-spawn" that we see in the ditches and ponds—often, it must be allowed, in places which a little more intelligence would have avoided.

It is profitable to pause to take a good look at this frog-spawn, for it illustrates a number of biological ideas, and perhaps we may be fortunate enough to see with a pocket-lens the eggs dividing into two, four, eight, and more cells, as if they were being cut by an invisible knife. Each egg in our common British frog *(Rana temporaria)* is about a tenth of an inch in diameter; it is almost entirely black, all but a small white lower pole; it is surrounded by a large sphere of non-living jelly, corresponding to the white of egg in a hen's egg; and there is no egg-shell. The whole mass, often of 2000 eggs, sinks at first, but afterwards floats freely.

Let us consider the biological significance of these spheres of jelly around the eggs, for it is very interesting to notice how they are justified on count after count, though they are non-living extrinsic investments. The spheres buoy up the eggs and at the same time obviate overcrowding. In the little chinks between the spheres there are often groups of green unicellular plants which liberate oxygen in the sunlight and use up the carbonic acid gas which the developing eggs produce—a most profitable association, a miniature illustration of the balance of Nature. But there is a fauna as well as a flora of frog-spawn, and the chinks are tenanted by small fry—such as water-fleas and rotifers—some of which eventually loosen the gelatinous envelopes, helping the larval-frogs to escape. Others, it must be admitted, seem to wait to devour. Once again, the envelopes of jelly are useful in lessening the risks of jostling—which might be fatal to the delicate embryos—when the wind raises waves in the pond, or when a water-hen or coot splashes in among the spawn. Moreover, the jelly seems to be unpalatable to most water animals, and it is so slippery that few birds can make anything of it. Finally, it may be that the clear spheres serve as so many greenhouses, enabling the ova to make the most of the sun's rays. All this illustrates the scientific view of Nature as an arena where efficiency in any form always counts.

About a fortnight or three weeks after the individual life began, that is, after fertilisation, the minute larvae are hatched from the delicate envelope of the ovum, and begin to wriggle about in the dissolving jelly. They are somewhat awkward-looking, half-made creatures at first, and when they emerge from the jelly they are mouthless, limbless, eyeless, and gill-less. They attach themselves, often in long rows, to water-weed, the adhesion being effected by a paired cement-gland below the position of the future mouth. A bulging on the ventral surface of the body indicates the position of the still unused remains of the legacy of yolk.

Soon after hatching three pairs of external gills grow out, the first much the largest, one upon each of the first three branchial arches. These are not comparable to the external gills of a young shark or skate, which are really elongated internal gills projecting through the gill-clefts. They are comparable to the true external gills seen in the young of some very archaic fishes, still living to-day, the *Polypterus* of the Gambia and some other African rivers, and two of the mud-fishes, *Protopterus* from Africa, and *Lepidosiren* from the Amazons. One or two days after hatching (in our common *Rana temporaria*) another important structure appears on the larva—the mouth is formed in the centre of a groove in front of the adhesive organs, and hundreds of small horny teeth are developed.

When the food-canal becomes open, four pairs of gill-clefts break out from the pharynx, and a gill-cover overlaps the first set of gills. These dwindle and are absorbed, their place being taken by a second set of gills supported by the hinder margin of the lower halves of four gill-arches. As these are enclosed in a gill-chamber and as they form a second set, it is natural to compare them to typical fish-gills. But they are really in the strict sense external, and they are certainly skin-covered. Each gill-chamber has at first its opening, but that of the right side joins with that of the left.

About a month after hatching the larval frog is in many ways fish-like: for instance, it has a two-chambered heart which drives impure blood to the gills, which are enclosed by a gill-cover. It swims by its laterally compressed tail, which shows a well developed unpaired fin, without fin-rays, however, which support the unpaired fins of fishes. *In a very general way* it may be said that the developing frog visibly climbs up its own genealogical tree. It is this general idea, indeed, of recapitulation that makes the study of the frog's life-history perennially interesting. It re-enacts the epoch-making colonisation of the dry land, and in many of its internal changes, *e.g.* in the making of the three-chambered heart, it probably re-enacts what took place very long ago—before the Coal-Measures were laid down in Britain—when amphibians evolved from a piscine stock. But what took the race long ages to accomplish is achieved by the individual in a few days,—a fact so familiar that we are apt to forget its marvellousness—the mystery of cumulative and condensed inheritance.

With the acquisition of a mouth the larva begins to feed eagerly, nibbling at plants in the water, and also eating animal food. As a consequence it grows, and the food-canal, in particular, becomes very long and coiled like a watch-spring. It is interesting to notice the relatively great length of the intestine during the predominantly vegetarian period, for it is usual in the animal kingdom to find a diet of vegetable food—which is somewhat slowly digested—associated with length of food-canal or with some equivalent of length. As the tail becomes stronger and the power of locomotion increases, the horse-shoe shaped adhesive organ is converted into two small discs which gradually disappear.

A new stage is marked by the appearance of the hind-limbs as minute projecting buds at the boundary between trunk and tail. Why should hind-limbs appear earlier than the fore-limbs, which are, moreover, much shorter? Investigation shows that they begin to develop at the same time—a fact which gives additional point to the question. The fore-limbs are delayed by the gill-cover, which does not impede the hind-limbs, and they eventually emerge, the left one through the "spiracle," the right one by a rupture. Perhaps we get some insight into the orderliness of developmental processes when we notice that the microscopic lashes or cilia which have hitherto covered the skin of the larva now disappear. In most cases, except as regards reproduction, what we may call "Animate Nature" (for shortness) is conspicuously *economical*.

After the appearance of the hind-legs, the larvae come often to the surface to breathe. They are learning to use their lungs, which have been slowly developing for some time as pockets projecting into the body-cavity from the under side of the gullet. The tadpoles are now about two months old, and in having lungs as well as gills they may be compared to the double-breathing Mud-Fishes or Dipnoi. As the lungs become established and functional, the gills dwindle, and an intricate series of internal changes leads from an essentially fish-like heart and circulation to the characteristic amphibian arrangements.

After a period of hearty feeding, with consequent increase in size and strength, the tadpole begins to show signs of approaching metamorphosis. It loses its appetite, it becomes much less energetic. The tail begins to break up internally, its muscles and other structures become disintegrated and dissolved, and most of the material is swept away in the blood stream to help in building up a better head. Wandering amoeboid cells, which are present in almost all animals except threadworms and lancelets, seem to play an important part in the extraordinary process of absorbing the tail, working like sappers and miners among the debris, dissolving some of the material, carrying some away. In certain respects what occurs is comparable to violent inflammation. It is like a pathological process which has become normal, and thus from watching tadpoles we get a glimpse of a deep-reaching theory of disease as "a perturbation which contains no elements essentially different from those of health, but elements presented in a different and less useful order." Often, at least, a disease implies a series of metabolic changes which are not in themselves in any way extraordinary—only they are out of place, out of time, and out of order.

One of the many careful observers of the annual wonder—the metamorphosis of the tadpole—gives the following terse statement of some of the more obvious changes: "The horny jaws are thrown off; the large frilled lips shrink up; the mouth loses its rounded suctorial form and becomes much wider; the tongue, previously small, increases considerably in size; the eyes, which as yet have been beneath the skin, become exposed."

As the tail shortens more and more, the tadpole, rapidly ceasing to be a tadpole, recovers its appetite and feeds greedily on animal matter, sometimes on its younger fellows. The abdomen shrinks, the stomach and liver enlarge, the intestine becomes relatively narrower and shorter. The tail is reduced to a short projecting stump, and, apart from this, the adult shape has been reached. Disinclination for a purely aquatic life becomes marked, and the young frogs clamber ashore. As they have lost all trace of gills, they are apt to drown in aquaria unless they have floating rafts to climb on to, or some other means of breathing dry air.

It is difficult to say which aspect of the development of tadpoles is most interesting. As we have seen, it is interesting in its main features as a modified recapitulation of that transition from aquatic to aerial respiration, from water to terra firma, which must have marked one of the most important epochs in the evolution of vertebrates.

But it is equally interesting to go into minute detail and notice that the young tadpole's small tongue has not much muscularity about it; that as the tongue increases in size the muscles also increase, but yet are quite unable to move the tongue, though perhaps [are] of some service in compressing glands; and that, as the metamorphosis is accomplished and the frogling hops ashore, the muscles of the tongue are at length strong enough to shoot out the tongue on the day-dreaming fly. The peculiar interest of this is that amphibians were the first animals to have a movable tongue, that of fishes being even worse than flabby, entirely non-muscular.

It is very interesting to consider in the same way the other momentous acquisitions made by the race of amphibians—such as fingers and toes, and the power of gripping things, vocal cords, and the power of speech—though how much they have to say in their extraordinary jabber no one knows.

Another interesting consideration is the variety of solutions that this one animal, the frog, offers to the problems of its life. Even in mathematics, we believe, there may be more than one solution to a problem, and everyone knows that this is true of the practical problems of human life. There is considerable variety in the solutions of the problems of *Brodwissenschaft,* though in strictness, we suppose, the fact of the matter is that the *conditions* of the typical problems are diverse, and therefore the solutions are diverse. But our point is this, that, to the two great problems of nutrition and respiration (if they are really *two,* for is not oxygen a kind of food?), the frog offers in the course of its life-history an unusual diversity of answer.

It will feed on its legacy of yolk, on unicellular algae, on the epidermis of aquatic plants, on the vegetable debris in the water, on animal matter by the way, on its own tail (of course in a sort of surreptitious phagocytic fashion), on its own brethren, on dead things in the water (tadpoles clean delicate skeletons beautifully), and, by and by, when it comes to its own, after a remarkable gustatory curriculum, it will feed on living insects and little else. Yet the way it feeds as an adult, *e.g.* on beetles much too large for it, is often far from saying much for its varied gastronomic education.

And again, as regards the fundamental problem of breathing, we find the newly hatched tadpole breathing through its skin in the old-fashioned manner of earthworm and leech; then follow in succession, the first set of external gills, the gill-clefts, the second set of external gills, which are usually called internal; then follows a period with gills and lungs together; then there is the transition to terrestrial life with pulmonary and (retained) cutaneous respiration; finally, in winter, the hibernating frog, retiring into the mud-fortresses of its remote ancestors, breathes by its skin only. . . .

Before leaving the tadpoles, interesting in so many ways, let us think over the year's life of the frog. Throughout the winter months the frogs lie near the pond, buried in the mud, mouth shut, nose shut, eyes shut, with the heart beating feebly, breathing through their skin, and eating nothing. The awakening in spring is followed immediately by pairing and egg-laying, and the aquatic juvenile life of the tadpoles occupies about three months. In summer there is a remarkable migration to the fields and meadows, and many hundreds of froglings, about the size of a first-finger nail, are seen on the march from the pond. The adults also migrate, and the meaning in both cases is the same—that they seek out places where insects abound. Of the many that go forth, only a remnant returns, for there is great mortality in the fields, where there are many physical risks and many alert enemies. The grass snake alone accounts for a good many in some parts of England. Those that escape—whether youngsters or old experienced hands—return to the pond in the autumn, and go into winter-quarters in the mud.

Julian Huxley

From *Essays in Popular Science* (1927)
"The Frog and Biology"

The frog is a too often despised animal,

> Whom there are few to praise,
> And very few to love.

It is small, defenceless, clammy, and rather ludicrous. But it is an organism, and one belonging to the same group of animals as man: it has, for the biologist, the merit of being abundant and easily procured, and that not only as adult but in all stages of its life-history. So it comes about that we know more about the frog, or rather about a number of processes in its life, than about the corresponding processes in any other single organism. "Know me, know my frog"— that is, I think, a legitimate adaptation of the old proverb for the biologist.

Do not imagine that I can even touch upon one-tenth of the biological problems packed into this little body. I only hope to show a few of the ways in which Frog may be profitably employed as text for biological sermons. You remember Herrick's lines on a praying child:

> Here a little child I stand,
> Heaving up my either hand,
> Cold as paddocks though they be,
> Here I lift them, Lord, to thee.

A paddock or puddock is a frog, as I expect you all know; and frogs are certainly cold. Many people lump together the coldness of snakes and lizards with that of frogs; but in reality the two are quite different. The snake has a smooth dry skin; the frog a clammy wet one. This clamminess has a meaning. The frog breathes in part through its skin, while snakes, like dogs and men, do not. For such cutaneous respiration the skin must be kept moist, for gases will not pass through a dry membrane, but the oxygen must actually enter into solution before it can be passed into the blood, and carbon dioxide stay in solution until it is in contact with the air if it is to be expelled from the body.

It might be supposed that it was a convenience to supplement one's lungs with one's skin as a breathing organ: and so no doubt in some ways it is—but on one condition. That condition is that the animal live in moist surroundings, for otherwise its skin, which must be wet and thin to exercise its respiratory function, would dry up.

Our frog, that is to say, is a land animal, if you like; but an animal of moist land only. It is a compromise between the aquatic and the terrestrial, emancipated only in part from the watery home of its ancestors.

The more we study the frog the more examples of such compromise do we find, notably in its skeleton, the architecture of its skull, the structure of its heart, and its life-history, in life-history perhaps most strikingly of all.

When I was a small boy, we used often to go and stay in the country with an aunt. Sunk in one corner of the garden was a large tank, perhaps six or eight feet deep, with vertical walls of cement, and more or less full of rain-water and duckweed. In the spring, this used to be occupied by quantities of frogs and toads. These set up a fine chorus of croaking (though why the soft pretty cooing of the toad should be called a croak I do not know), and we often came down to look at them as they lay suspended in the water, eyes and nose just emerging, forelegs spread, and body and hind-legs trailing diagonally down.

Somehow or other we discovered the fact that they liked to scramble on to any floating piece of wood in the water; and we used to spend hours giving the frogs a ride. We took a straight log from the woodpile, and, quite illegally, hurled it in the water. The frogs (I can see them now) swam from all sides and climbed out upon it. Then with long hazel sticks we propelled the log from one end of the tank to the other. It would rotate a little, and spill the less skilful of its passengers; and it was our amusement to see them try to mount again in mid-course. Some of them, however, were very expert, and would stick on for long journeys.

I do not know why this pleased us so much, but it did; and it remains as one of the vivid memories of the place. However, to discuss that would take us into the biology of other organisms than frogs!

The frogs came to the tank in the spring to breed. But what we boys did not think of was the fact that although they could easily enough get in, they could not get out: and that therefore none of their offspring would ever be able to come out on land, but would perish with their parents in the water at the base of those unscaleable walls.

Why did the frogs come to breed in the tank? They came because it was the nearest piece of water, and because frogs must breed in water. Their eggs are fertilised outside the body, and will dry up if not in water; and they develop into tadpoles, which can only breathe and feed and move in water. Not merely this, but in many essentials of structure the tadpole *is* a fish—in the arrangement of its gills and the blood-vessels that supply them, in its heart, the fin along its tail, and the special sense-organs for perceiving low-frequency vibrations in water which, like a herring or any other fish, it carries on a "lateral line" along its flank.

There can be no doubt whatever from all the different lines of evidence which we have at command that frogs are actually descended from fish, even if from none of the common types of fish with which we are familiar; and that is to say that they once spent the whole of their life in the water. Now the difficulties attendant upon emancipation from water into air press much harder upon the early than upon the late stages of development. The act of fertilisation itself must be accomplished in a fluid medium. The delicate tissues of the egg and embryo must be bathed in fluid. Life in a fluid with weight supported by displacement is far simpler for the early stages of the animal when the skeleton has not yet been formed.

These difficulties can be and have been surmounted by all the highest land vertebrates. Fertilisation becomes internal instead of external. The great biological invention, the amnion, came into existence—an overarching membrane grown by the embryo for its own protection, enclosing it in a resistant water-cushion, or, as one writer picturesquely puts it, enabling the embryo to live and develop within its own private pond. The private pond in its turn is protected in a hard shell, tight packed with weeks' or even months' supply of provisions; or else is still more efficiently sheltered and victualled within the body of the mother. Only so do vertebrates become truly terrestrial, for only so do they become independent of water at all stages of their life-history.

This emancipation from the water was one of the few large steps in the progress of vertebrates from some primitive lamprey-like form. First the early acquisition of limbs, jaws, teeth, and bone; then this conquest of the land; then the stabilising of the animal's temperature; and finally the development of mind to its highest pitch. In a sense, the transition to terrestrial existence was the most abrupt of all, save the transition from irrational to reasoned life: so it is natural that the adjustment should need time, and that many makeshifts should come into being. . . .

Very well, you may say: but if evolution has brought the frog-type into being, why did it not bring it to a still higher level of existence? If progressive change is one of the laws of evolution, why is it that there are species and groups which do not progress, but stand still, resting on their evolutionary laurels? The question is a fundamental one, and one which has exercised many minds. But the answer, I believe, is not so difficult as at first sight appears.

The organic world develops in a constant condition of what we may symbolise as pressure—the pressure of the inorganic environment, the pressure of enemies, of competition for food and space and mates.

Anything which will give an organism an advantage in the struggle against environment or enemies or competitors of its own blood will reduce this biological pressure. This is especially true where a real progressive variation is concerned, where some character is developed which enables a species to invade hitherto untrodden fields—in other words to invade a low-pressure area; or when the pressure is reduced through the development of some improvement of general organisation—a procedure comparable to the reorganisation of a business from within.

Let us now take a concrete case. Why, when the amphibian type gave rise to the reptilian, did not all amphibia become transformed, or why did those which failed to do so not become extinct under the stress of the new competition from the improved type? The answer is that the great advantage gained by the reptiles lay in the opening up of a whole new area to vertebrate colonisation. In the old area already colonised, their advantage, if any, would not be so great; and in parts of it (for instance the swampier regions) the old type filled the bill so perfectly that the advantage lay on their side. Precisely the same considerations hold good for the inventions of man. The wheeled horse-drawn vehicle ousted the riding and pack animal as the chief means of transport, to be ousted in its turn by the mechanically-propelled vehicle. But there are still particular places, and particular kinds of transport, in which the horse-drawn vehicle still has advantages over the motor, and still other regions in which any wheeled vehicle is at a disadvantage. As a result the old types do survive, though restricted, alongside the new. So the largest and the most progressive types of amphibia perished in competition with the reptiles; but there is a niche well filled by the amphibian type, and competition is not enough to oust the inconspicuous and specialised forms that remain in it. . . .

. . . The frog compromises with its aquatic past by existing in the form of a tadpole through the early part of its life. To extricate itself from this compromise, it requires a regular *volte-face;* and this *volte-face* is the brusque transformation of tadpole into froglet which we call metamorphosis.

In this apparently simple phenomenon which occurs, unregarded, uncounted millions of times before our eyes each summer, we have in reality a most complex and delicately-adjusted set of processes. Think for a moment of all that it involves. The limbless tadpole must grow limbs, but at the same time must rid itself of tail. The machinery for water-breathing must be discarded, and replaced by one for air-breathing. The skin must be altered to fit it for land life: *inter alia,* means must be provided to keep it moist against the danger of desiccation. In the water, its weight was negligible: in air, that will no longer be so. Mr. Wells and other scientific romancers have made great play with the difference in weight which would be experienced on the moon by an inhabitant of earth: but the extent of the change of weight undergone by every tadpole which successfully turns into a frog is even greater.

Then the digestive system must be remodelled to suit a flesh diet instead of one predominantly vegetable: the horny jaws thrown off, true teeth grown, the intestine's long coils shortened, the liver and pancreas altered.

All these and other changes must all conspire to take effect at the same time. . . .

I have wandered far afield from the frogs of my boyish memories, trapped by their own instincts in the sunken tank; but I am now going to return to them for a fresh start.

What first attracted our attention to them was the noise they made, a chorus of croaking, harsh and raucous from the frogs, soft and musical from the toads. That, as you know, is all spring music; but it is music which is found in almost every species of frog or toad in existence. In our northern climate, this frog-music is a quiet little affair; but in more southern countries it provides one of the memorable sounds of nature. In France they have five or six species of batrachians against our three; and in the spring in marshy parts of the country they sing to such purpose that under the *ancien régime* one of the tasks imposed upon the peasants was to beat the waters of the ponds at night so that the frogs would hush and let the nobility slumber. In the southern States of America there are grass-frogs, and leopard-frogs, and bullfrogs, and toads, and tree toads; and the voices of some of them are far more powerful than ours. In the spring a few wet days will fill the prairie pools, and the pools will fill with toads and frogs, and then the evenings and nights will be full of sound—a continuous sound, rising and falling like the sound of the sea. For three years I lived in Texas, and when I left, I think that manifestation of Nature which I most missed was the sound of the frogs in spring. Others too who have lived in the south have told me that they felt the same.

What is this sound and its meaning? It is produced with the aid of large resonators, in the shape of distensible pockets of skin under the throat or on either side of the neck. This is in the males only, for the females do not croak. A small male tree frog croaking, with expanded pouches bigger than his own head, is a never-to-be-forgotten spectacle.

And the purpose and function of the croaking? The function seems to be simply that of recognition between the sexes, the females being guided to their mates by the sound. "The frog he would a-wooing go—'Heigho!' said Anthony Rowley"—and "croak, croak" says the frog himself.

In this respect too the frog is in a sense at an intermediate level of animal evolution. The lowest multicellular animals shed their gametes or reproductive cells free into the water, leaving chance to do the rest; at most, a certain synchronisation of the shedding on the part of a large number of individuals is achieved in some cases, probably by some chemical stimulus of one individual on the rest, in others apparently altogether under the regulation of outer influences such as the seasons or the tides. Before the evolution of complex sense-organs and still more of brains behind those sense-organs, anything more elaborate would have been useless and impossible.

At the other extreme we have creatures like birds with perfected eyes and ears and brains so elaborate as to permit of a complex emotional life. Reproduction has become so much more complex that fertilisation must be internal, and an act of pairing is therefore essential. Furthermore, no longer is the liberation of the gametes a mere reflex action, nor yet a simple instinct, but is under the control of the highest centres, and demands the proper adjustment of the emotional system before it can be carried out. One of the main functions of the strange and beautiful dances and displays which are to be seen in birds at courting-time is undoubtedly to tune the emotions of the two birds of a pair, male and female, to the same key simultaneously and so facilitate mating. Even among birds, however, evolution has here and there gone on to a higher stage. In grebes and herons and other birds in which the sexes are alike and both indulge in elaborate display, the display often seems to be performed for its own sake, without special stimulative function. The only function it can then be supposed to keep is that of linking the pair together throughout the breeding season by emotional bonds; this may well have its biological value, since in all such species, both sexes help in incubation and in attending to the young, and anything which binds the pair more closely together for the good of the family is a racial blessing. This emancipation of courtship and display from a merely stimulative function has of course in man reached up higher on to new levels, and out wider so as to commingle with other parts of the emotional life—higher and wider than in any other creature.

But of all this in the frog there is no trace. The actual mating is still a merely or essentially physiological process, not a psychological one. However, the sexes must be brought together with as little waste as possible; brain and sense-organs have been developed to a comparatively high pitch to subserve the general functions of catching prey and escaping from enemies—they can be brought under call to supply this new demand.

In part, the male frog's croaking is like the male spider's dancing or still better the male fiddler-crab's statuesque poses with enormous claw uplifted—it is an advertisement of his maleness. But in part it is an advertisement of the breeding grounds, a boasting of choice real estate; and in this frogs foreshadow song-birds, whose song is a sky-sign read by other birds according to their sex—to other males, a warning to trespassers, to the females, an advertisement of home and husband.

In any event, the amphibian's croaking is interesting from yet another biological aspect. It was the first vocal music that life brought to birth. No doubt insects had been deliberately producing their chirpings, trillings, and hummings before fish ever left the water; but they are instrumentalists all, making zithers of their legs or fiddles of their wings.

The land vertebrates offered life two new possibilities of sound-production—the pipe with reed, in the shape of the windpipe and vocal chords; and, in the mouth, a cavity communicating with the pipe and alterable in shape. The first gave the possibility of varying pitch, the second of altering the type and form of the sounds: and so it is sober truth to say that only through some humble beginning such as that which the croaking frog gives us in spring, could bird song and human song and speech have come into existence. . . .

So the race of frogs continues, fixed it would seem irrevocably in an evolutionary half-way house, a peculiar and specialised manifestation of life, risen far above life's primitive estate, and exhibiting many adumbrations of her higher achievements. Each year they mate, and generate new frog-lives, and die. Their whole existence is delicately regulated by the never-ending pressure of circumstance, to the world around, and attuned to what appears to be its only end—its own adequate performance.

Its only end, but not its only function: for each organism performs many functions in respect of other organisms, whether as food for this one, or in keeping this other up to full pitch by constant pursuit. And here today the frog has even been performing a function in the sphere of the intellect, in providing a text for my biological sermon!

What we know of modern physics assures us that to separate any particle of matter from the rest of the universe can be only either an intellectual trick or a practical approximation. Matter is in the ultimate analysis nothing but a vast number of centres of activity. The activity diminishes as we pass away from the centre, like the wave from a stone's splash; it increases as we pass toward the centre until finally the forces at work are enormous, the remaining core becomes what we call impenetrable, and we speak of it as a material unit. But in reality the influences of the activities of all the units interpenetrate in the one cosmos.

So it is in the mental kingdom. It is impossible to pose a question without a reverberation which raises up further questions, one after the other, till, if we are willing in the pursuit, we find we have started a stir in the whole organism of knowledge.

Let my brief and incomplete words to-day at least serve as a reminder of this fact, and as an appeal to teachers not to allow themselves or their pupils to become so smothered with facts and details and the requirements of examining boards that they lose sight of this reality, and with it the light of intellectual day.

LOREN EISELEY

From *The Star Thrower* (1978)
"The Dance of the Frogs"

He was a member of the Explorers Club, and he had never been outside the state of Pennsylvania. Some of us who were world travelers used to smile a little about that, even though we knew his scientific reputation had been, at one time, great. It is always the way of youth to smile. I used to think of myself as something of an adventurer, but the time came when I realized that old Albert Dreyer, huddling with his drink in the shadows close to the fire, had journeyed farther into the Country of Terror than any of us would ever go, God willing, and emerge alive.

He was a morose and aging man, without family and without intimates. His membership in the club dated back into the decades when he was a zoologist famous for his remarkable experiments upon amphibians—he had recovered and actually produced the adult stage of the Mexican axolotl, as well as achieving remarkable tissue transplants in salamanders. The club had been flattered to have him then, travel or no travel, but the end was not fortunate. The brilliant scientist had become the misanthrope; the achievement lay all in the past, and Albert Dreyer kept to his solitary room, his solitary drink, and his accustomed spot by the fire.

The reason I came to hear his story was an odd one. I had been north that year, and the club had asked me to give a little talk on the religious beliefs of the Indians of the northern forest, the Naskapi of Labrador. I had long been a student of the strange melange of superstition and woodland wisdom that makes up the religious life of the nature peoples. Moreover, I had come to know something of the strange similarities of the "shaking tent rite" to the phenomena of the modern medium's cabinet.

"The special tent with its entranced occupant is no different from the cabinet," I contended. "The only difference is the type of voices that emerge. Many of the physical phenomena are identical—the movement of powerful forces shaking the conical hut, objects thrown, all this is familiar to Western psychical science. What is different are the voices projected. Here they are the cries of animals, the voices from the swamp and the mountain—the solitary elementals before whom the primitive man stands in awe, and from whom he begs sustenance. Here the game lords reign supreme; man himself is voiceless."

A low, halting query reached me from the back of the room. I was startled, even in the midst of my discussion, to note that it was Dreyer.

"And the game lords, what are they?"

"Each species of animal is supposed to have gigantic leaders of more than normal size," I explained. "These beings are the immaterial controllers of that particular type of animal. Legend about them is confused. Sometimes they partake of human qualities, will and intelligence, but they are of animal shape. They control the movements of game, and thus their favor may mean life or death to man."

"Are they visible?" Again Dreyer's low, troubled voice came from the back of the room.

"Native belief has it that they can be seen on rare occasions," I answered. "In a sense they remind one of the concept of the archetypes, the originals behind the petty show of our small, transitory existence. They are the immortal renewers of substance—the force behind and above animate nature."

"Do they dance?" persisted Dreyer.

At this I grew nettled. Old Dreyer in a heckling mood was something new. "I cannot answer that question," I said acidly. "My informants failed to elaborate upon it. But they believe implicitly in these monstrous beings, talk to and propitiate them. It is their voices that emerge from the shaking tent."

"The Indians believe it," pursued old Dreyer relentlessly, "but do *you* believe it?"

"My dear fellow"—I shrugged and glanced at the smiling audience—"I have seen many strange things, many puzzling things, but I am a scientist." Dreyer made a contemptuous sound in his throat and went back to the shadow out of which he had crept in his interest. The talk was over. I headed for the bar.

The evening passed. Men drifted homeward or went to their rooms. I had been a year in the woods and hungered for voices and companionship. Finally, however, I sat alone with my glass, a little mellow, perhaps, enjoying the warmth of the fire and remembering the blue snowfields of the North as they should be remembered—in the comfort of warm rooms.

I think an hour must have passed. The club was silent except for the ticking of an antiquated clock on the mantel and small night noises from the street. I must have drowsed. At all events it was some time before I grew aware that a chair had been drawn up opposite me. I started.

"A damp night," I said.

"Foggy," said the man in the shadow musingly. "But not too foggy. They like it that way."

"Eh?" I said. I knew immediately it was Dreyer speaking. Maybe I had missed something; on second thought, maybe not.

"And spring," he said. "Spring. That's part of it. God knows why, of course, but we feel it, why shouldn't they? And more intensely."

"Look—" I said. "I guess—" The old man was more human than I thought. He reached out and touched my knee with the hand that he always kept a glove over—burn, we used to speculate—and smiled softly.

"You don't know what I'm talking about," he finished for me. "And, besides, I ruffled your feelings earlier in the evening. You must forgive me. You touched on an interest of mine, and I was perhaps overeager. I did not intend to give the appearance of heckling. It was only that . . ."

"Of course," I said. "Of course." Such a confession from Dreyer was astounding. The man might be ill. I rang for a drink and decided to shift the conversation to a safer topic, more appropriate to a scholar.

"Frogs," I said desperately, like any young ass in a china shop. "Always admired your experiments. Frogs. Yes."

I give the old man credit. He took the drink and held it up and looked at me across the rim. There was a faint stir of sardonic humor in his eyes.

"Frogs, no," he said, "or maybe yes. I've never been quite sure. Maybe yes. But there was no time to decide properly." The humor faded out of his eyes. "Maybe I should have let go," he said. "It was what they wanted. There's no doubting that at all, but it came too quick for me. What would you have done?"

"I don't know," I said honestly enough and pinched myself.

"You had better know," said Albert Dreyer severely, "if you're planning to become an investigator of primitive religions. Or even not. I wasn't, you know, and the things came to me just when I least suspected—But I forget, you don't believe in them."

He shrugged and half rose, and for the first time, really, I saw the black-gloved hand and the haunted face of Albert Dreyer and knew in my heart the things he had stood for in science. I got up then, as a young man in the presence of his betters should get up, and I said, and I meant it, every word: "Please, Dr. Dreyer, sit down and tell me. I'm too young to be saying what I believe or don't believe in at all. I'd be obliged if you'd tell me."

Just at that moment a strange, wonderful dignity shone out of the countenance of Albert Dreyer, and I knew the man he was. He bowed and sat down, and there were no longer the barriers of age and youthful ego between us. There were just two men under a lamp, and around them a great waiting silence. Out to the ends of the universe, I thought fleetingly, that's the way with man and his lamps. One has to huddle in, there's so little light and so much space. One—

"It could happen to anyone," said Albert Dreyer. "And especially in the spring. Remember that. And all I did was to skip. Just a few feet, mark you, but I skipped. Remember that, too.

"You wouldn't remember the place at all. At least not as it was then." He paused and shook the ice in his glass and spoke more easily.

"It was a road that came out finally in a marsh along the Schuylkill River. Probably all industrial now. But I had a little house out there with a laboratory thrown in. It was convenient to the marsh, and that helped me with my studies of amphibia. Moreover, it was a wild, lonely road, and I wanted solitude. It is always the demand of the naturalist. You understand that?"

"Of course," I said. I knew he had gone there, after the death of his young wife, in grief and loneliness and despair. He was not a man to mention such things. "It is best for the naturalist," I agreed.

"Exactly. My best work was done there." He held up his black-gloved hand and glanced at it meditatively. "The work on the axolotl, newt neoteny. I worked hard. I had"—he hesitated—"things to forget. There were times when I worked all night. Or diverted myself, while waiting the result of an experiment, by midnight walks. It was a strange road. Wild all right, but paved and close enough to the city that there were occasional street lamps. All uphill and downhill, with bits of forest leaning in over it, till you walked in a tunnel of trees. Then suddenly you were in the marsh, and the road ended at an old, unused wharf.

"A place to be alone. A place to walk and think. A place for shadows to stretch ahead of you from one dim lamp to another and spring back as you reached the next. I have seen them get tall, tall, but never like that night. It was like a road into space."

"Cold?" I asked.

"No. I shouldn't have said 'space.' It gives the wrong effect. Not cold. Spring. Frog time. The first warmth, and the leaves coming. A little fog in the hollows. The way they like it then in the wet leaves and bogs. No moon, though; secretive and dark, with just those street lamps wandered out from the town. I often wondered what graft had brought them there. They shone on nothing—except my walks at midnight and the journeys of toads, but still . . ."

"Yes?" I prompted, as he paused.

"I was just thinking. The web of things. A politician in town gets a rake-off for selling useless lights on a useless road. If it hadn't been for that, I might not have seen them. I might not even have skipped. Or, if I had, the effect—How can you tell about such things afterwards? Was the effect heightened? Did it magnify their power? Who is to say?"

"The skip?" I said, trying to keep things casual. "I don't understand. You mean, just skipping? Jumping?"

Something like a twinkle came into his eyes for a moment. "Just that," he said. "No more. You are a young man. Impulsive? You should understand."

"I'm afraid—" I began to counter.

"But of course," he cried pleasantly. "I forget. You were not there. So how could I expect you to feel or know about this skipping. Look, look at me now. A sober man, eh?"

I nodded. "Dignified," I said cautiously.

"Very well. But, young man, there is a time to skip. On country roads in the spring. It is not necessary that there be girls. You will skip without them. You will skip because something within you knows the time—frog time. Then you will skip."

"Then I will skip," I repeated, hypnotized. Mad or not, there was a force in Albert Dreyer. Even there under the club lights, the night damp of an unused road began to gather.

"It was a late spring," he said. "Fog and mist in those hollows in a way I had never seen before. And frogs, of course. Thousands of them, and twenty species, trilling, gurgling, and grunting in as many keys. The beautiful keen silver piping of spring peepers arousing as the last ice leaves the ponds—if you have heard that after a long winter alone, you will never forget it." He paused and leaned forward, listening with such an intent inner ear that one could almost hear that far-off silver piping from the wet meadows of the man's forgotten years.

I rattled my glass uneasily, and his eyes came back to me. "They come out then," he said more calmly. "All amphibia have to return to the water for mating and egg laying. Even toads will hop miles across country to streams and waterways. You don't see them unless you go out at night in the right places as I did, but that night—

"Well, it was unusual, put it that way, as an understatement. It was late, and the creatures seemed to know it. You could feel the forces of mighty and archaic life welling up from the very ground. The water was pulling them—not water as we know it, but the mother, the ancient life force, the thing that made us in the days of creation, and that lurks around us still, unnoticed in our sterile cities.

"I was no different from any other young fool coming home on a spring night, except that as a student of life, and of amphibia in particular, I was, shall we say, more aware of the creatures. I had performed experiments"—the black glove gestured before my eyes. "I was, as it proved, susceptible.

"It began on that lost stretch of roadway leading to the river, and it began simply enough. All around, under the street lamps, I saw little frogs and big frogs hopping steadily toward the river. They were going in my direction.

"At that time I had my whimsies, and I was spry enough to feel the tug of that great movement. I joined them. There was no mystery about it. I simply began to skip, to skip gaily, and enjoy the great bobbing shadow I created as I passed onward with that leaping host all headed for the river.

"Now skipping along a wet pavement in spring is infectious, particularly going downhill, as we were. The impulse to take mightier leaps, to soar farther, increases progressively. The madness worked into me. I bounded till my lungs labored, and my shadow, at first my own shadow, bounded and labored with me.

"It was only midway in my flight that I began to grow conscious that I was not alone. The feeling was not strong at first. Normally a sober pedestrian, I was ecstatically preoccupied with the discovery of latent stores of energy and agility which I had not suspected in my subdued existence.

"It was only as we passed under a street lamp that I noticed, beside my own bobbing shadow, another great, leaping grotesquerie that had an uncanny suggestion of the frog world about it. The shocking aspect of the thing lay in its size, and the fact that, judging from the shadow, it was soaring higher and more gaily than myself.

"'Very well,' you will say"—and here Dreyer paused and looked at me tolerantly—"'Why didn't you turn around? That would be the scientific thing to do.'

"It would be the scientific thing to do, young man, but let me tell you it is not done—not on an empty road at midnight—not when the shadow is already beside your shadow and is joined by another, and then another.

"No, you do not pause. You look neither to left nor right, for fear of what you might see there. Instead, you dance on madly, hopelessly. Plunging higher, higher, in the hope the shadows will be left behind, or prove to be only leaves dancing, when you reach the next street light. Or that whatever had joined you in this midnight bacchanal will take some other pathway and depart.

"You do not look—you cannot look—because to do so is to destroy the universe in which we move and exist and have our transient being. You dare not look, because, beside the shadows, there now comes to your ears the loose-limbed slap of giant batrachian feet, not loud, not loud at all, but there, definitely there, behind you at your shoulder, plunging with the utter madness of spring, their rhythm entering your bones until you too are hurtling upward in some gigantic ecstasy that it is not given to mere flesh and blood to long endure.

"I was part of it, part of some mad dance of the elementals behind the show of things. Perhaps in that night of archaic and elemental passion, that festival of the wetlands, my careless hopping passage under the street lights had called them, attracted their attention, brought them leaping down some fourth-dimensional roadway into the world of time.

"Do not suppose for a single moment I thought so coherently then. My lungs were bursting, my physical self exhausted, but I sprang, I hurtled, I flung myself onward in a company I could not see, that never outpaced me, but that swept me with the mighty ecstasies of a thousand springs, and that bore me onward exultantly past my own doorstep, toward the river, toward some pathway long forgotten, toward some unforgettable destination in the wetlands and the spring.

"Even as I leaped, I was changing. It was this, I think, that stirred the last remnants of human fear and human caution that I still possessed. My will was in abeyance; I could not stop. Furthermore, certain sensations, hypnotic or otherwise, suggested to me that my own physical shape was modifying, or about to change. I was leaping with a growing ease. I was—

"It was just then that the wharf lights began to show. We were approaching the end of the road, and the road, as I have said, ended in the river. It was this, I suppose, that startled me back into some semblance of human terror. Man is a land animal. He does not willingly plunge off wharfs at midnight in the monstrous company of amphibious shadows.

"Nevertheless their power held me. We pounded madly toward the wharf, and under the light that hung above it, and the beam that made a cross. Part of me struggled to stop, and part of me hurtled on. But in that final frenzy of terror before the water below engulfed me I shrieked, *'Help! In the name of God, help me! In the name of Jesus, stop!'*"

Dreyer paused and drew in his chair a little closer under the light. Then he went on steadily.

"I was not, I suppose, a particularly religious man, and the cries merely revealed the extremity of my terror. Nevertheless this is a strange thing, and whether it involves the crossed beam, or the appeal to a Christian deity, I will not attempt to answer.

"In one electric instant, however, I was free. It was like the release from demoniac possession. One moment I was leaping in an inhuman company of elder things, and the next moment I was a badly shaken human being on a wharf. Strangest of all, perhaps, was the sudden silence of that midnight hour. I looked down in the circle of the arc light, and there by my feet hopped feebly some tiny froglets of the great migration. There was nothing impressive about them, but you will understand that I drew back in revulsion. I have never been able to handle them for research since. My work is in the past."

He paused and drank, and then, seeing perhaps some lingering doubt and confusion in my eyes, held up his black-gloved hand and deliberately pinched off the glove.

A man should not do that to another man without warning, but I suppose he felt I demanded some proof. I turned my eyes away. One does not like a webbed batrachian hand on a human being.

As I rose embarrassedly, his voice came up to me from the depths of the chair.

"It is not the hand," Dreyer said. "It is the question of choice. Perhaps I was a coward, and ill prepared. Perhaps"—his voice searched uneasily among his memories—"perhaps I should have taken them and that springtime without question. Perhaps I should have trusted them and hopped onward. Who knows? They were gay enough, at least."

He sighed and set down his glass and stared so intently into empty space that, seeing I was forgotten, I tiptoed quietly away.

STEPHEN JAY GOULD

From *Bully for Brontosaurus: Reflections in Natural History* (1991)
"Here Goes Nothing"

Goliath paid the highest of prices to learn the most elementary of lessons—thou shalt not judge intrinsic quality by external appearance. When the giant first saw David, "he disdained him: for he was but a youth, and ruddy, and of a fair countenance" (I Sam. 17:42). Saul had been similarly unimpressed when David presented himself as an opponent for Goliath and savior of Israel. Saul doubted out loud: "for thou art but a youth, and he a man of war from his youth" (I Sam. 17:33). But David persuaded Saul by telling him that actions speak louder than appearances—for David, as a young shepherd, had rescued a lamb from a predatory lion: "I went out after him, and smote him, and delivered it out of his mouth" (I Sam. 17:35).

This old tale presents a double entendre to introduce this essay—first as a preface to my opening story about a famous insight deceptively clothed in drab appearance; and second as a quirky lead to the body of this essay, a tale of animals that really do deliver from their mouths: *Rheobatrachus silus,* an

Australian frog that swallows its fertilized eggs, broods tadpoles in its stomach, and gives birth to young frogs through its mouth.

Henry Walter Bates landed at Pará (now Belém), Brazil, near the mouth of the Amazon, in 1848. He arrived with Alfred Russel Wallace, who had suggested the trip to tropical jungles, arguing that a direct study of nature at her richest might elucidate the origin of species and also provide many fine specimens for sale. Wallace returned to England in 1852, but Bates remained for eleven years, collecting nearly 8,000 new species (mostly insects) and exploring the entire Amazon valley.

In 1863, Bates published his two-volume classic, perhaps the greatest work of nineteenth-century natural history and travel, *The Naturalist on the River Amazons*. But two years earlier, Bates had hidden his most exciting discovery in a technical paper with a disarmingly pedestrian title: "Contributions to an Insect Fauna of the Amazon Valley," published in the *Transactions of the Linnaean Society*. The reviewer of Bates's paper (*Natural History Reviews*, 1863, pp. 219–224) lauded Bates's insight but lamented the ill-chosen label: "From its unpretending and somewhat indefinite title," he wrote, "we fear [that Bates's work] may be overlooked in the ever-flowing rush of scientific literature." The reviewer therefore sought to rescue Bates from his own modesty by providing a bit of publicity for the discovery. Fortunately, he had sufficient oomph to give Bates a good send-off. The reviewer was Charles Darwin, and he added a section on Bates's insight to the last edition of the *Origin of Species*.

Bates had discovered and correctly explained the major style of protective mimicry in animals. In Batesian mimicry (for the phenomenon now bears his name), uncommon and tasty animals (the mimics) gain protection by evolving uncanny resemblance to abundant and foul-tasting creatures (the models) that predators learn to avoid. The viceroy butterfly is a dead ringer for the monarch, which, as a caterpillar, consumes enough noxious poisons from its favored plant foods to sicken any untutored bird. (Vomiting birds have become a cliche of natural history films. Once afflicted, twice shy, as the old saying goes. The tale may be more than twice told, but many cognoscenti do not realize that the viceroy's name memorializes its mimicry—for this butterfly is the surrogate, or vice-king, to the ruler, or monarch, itself.)

Darwin delighted in Bates's discovery because he viewed mimicry as such a fine demonstration of evolution in action. Creationism, Darwin consistently argued, cannot be disproved directly because it claims to explain everything. Creationism becomes impervious to test and, therefore, useless to science. Evolutionists must proceed by showing that any creationist explanation becomes a *reductio ad absurdum* by twists of illogic and special pleading required to preserve the idea of God's unalterable will in the face of evidence for historical change.

In his review of Bates's paper, Darwin emphasizes that creationists must explain the precision of duplicity by mimics as a simple act of divine construction—"they were thus clothed from the hour of their creation," he writes. Such a claim, Darwin then argues, is even worse than wrong because it stymies science by providing no possible test for truth or falsity—it is an argument "made at the expense of putting an effectual bar to all further inquiry." Darwin then presents his *reductio ad absurdum,* showing that any fair-minded person must view mimicry as a product of historical change.

Creationists had made a central distinction between true species, or entities created by God, and mere varieties, or products of small changes permitted within a created type (breeds of dogs or strains of wheat, for example). But Bates had shown that some mimics are true species and others only varieties of species that lack mimetic features in regions not inhabited by the model. Would God have created some mimics from the dust of the earth but allowed others to reach their precision by limited natural selection within the confines of a created type? Is it not more reasonable to propose that mimicking species began as varieties and then evolved further to become separate entities? And much worse for creationists: Bates had shown that some mimicking species resemble models that are only varieties. Would God have created a mimic from scratch to resemble another form that evolved (in strictly limited fashion) to its current state? God may work in strange ways, his wonders to perform—but would he really so tax our credulity? The historical explanation makes so much more sense.

But if mimicry became a source of delight for Darwin, it also presented a serious problem. We may easily grasp the necessity for a historical account. We may understand *how* the system works once all its elements develop, but *why* does this process of mimicry ever begin? What starts it off, and what propels it forward? Why, in Darwin's words, "to the perplexity of naturalists, has nature condescended to the tricks of the stage?" More specifically: Any butterfly mimic, in the rich faunas of the Amazon valley, shares its space with many potential models. Why does a mimic converge upon one particular model? We can understand how natural selection might perfect a resemblance already well established, but what begins the process along one of many potential pathways—especially since we can scarcely imagine that a 1 or 2 percent resemblance to a model provides much, if any, advantage for a mimic. This old dilemma in evolutionary theory even has a name in the jargon of my profession—the problem of the "incipient stages of useful structures." Darwin had a good answer for mimicry, and I will return to it after a long story about frogs—the central subject of this essay and another illustration of the same principle that Darwin established to resolve the dilemma of incipient stages.

We remember Darwin's *Beagle* voyage primarily for the big and spectacular animals that he discovered or studied: the fossil *Toxodon* and the giant Galápagos tortoises. But many small creatures, though less celebrated, brought enormous scientific reward—among them a Chilean frog appropriately named *Rhinoderma darwini*. Most frogs lay their eggs in water and then allow the tadpoles to make their own way, but many species have evolved various styles of parental care, and the range of these adaptations extols nature's unity in diversity.

In *R. darwini,* males ingest the fertilized eggs and brood them in the large throat pouches usually reserved for an earlier act of courtship—the incessant croaking that defines territory and attracts females. Up to fifteen young may fill the pouch, puffing out all along the father's ventral (lower) surface and compressing the vital organs above. G. B. Howes ended his classic account of this curious life-style (*Proceedings of the Zoological Society of London,* 1888) with a charming anthropomorphism. Previous students of *Rhinoderma,* he noted, had supposed that the male does not feed while carrying his young. But Howes dissected a breeding male and found its stomach full of beetles and flies and its large intestine clogged with "excreta like that of a normal individual." He concluded, with an almost palpable sigh of relief, "that this extraordinary paternal instinct does not lead up to that self-abnegation" postulated by previous authors.

But nature consistently frustrates our attempts to read intrinsic solicitude into her ways. In November 1973, two Australian scientists discovered a form of parental care that must preclude feeding, for these frogs brood their young in their stomachs and then give birth through their mouths. And we can scarcely imagine that a single organ acts as a nurturing uterus and a site of acid digestion at the same time.

Rheobatrachus silus, a small aquatic frog living under stones or in rock pools of shallow streams and rills in a small area of southeast Queensland, was first discovered and described in 1973. Later that year, C. J. Corben and G. J. Ingram of Brisbane attempted to transfer a specimen from one aquarium to another. To their astonishment, it "rose to the surface of the water and, after compression of the lateral body muscles, propulsively ejected from the mouth six living tadpoles" (from the original description published by Corben, Ingram, and M. J. Tyler in 1974).[1] They initially assumed, from their knowledge of *Rhinoderma,* that their brooder was a male rearing young in its throat pouch. Eighteen days later, they found a young frog swimming beside its parent; two days later, a further pair emerged unobserved in the night. At that point, they decided (as the euphemism goes) to "sacrifice" their golden goose. But the parent, when grasped, "ejected by propulsive vomiting eight juveniles in the space of no more than two seconds. Over the next few minutes a further five juveniles were ejected." They then dissected the parent and received their biggest surprise. The frog had no vocal sac. It was a female with "a very large, thin-walled, dilated stomach"—the obvious home of the next generation.

Natural birth had not yet been observed in *Rheobatrachus*. All young had either emerged unobserved or been vomited forth as a violent reaction after handling. The first young had greeted the outside world prematurely as tadpoles (since development clearly proceeds all the way to froghood in the mother's stomach, as later births demonstrated).

Art then frustrated nature, and a second observation also failed to resolve the mode of natural birth. In January 1978, a pregnant female was shipped express airfreight from Brisbane to Adelaide for observation. But the poor frog was—yes, you guessed it—"delayed" by an industrial dispute. The mother, still hanging on, eventually arrived surrounded by twenty-one dead young; a twenty-second frog remained in her stomach upon dissection. Finally, in 1979, K. R. McDonald and D. B. Carter successfully transported two pregnant females to Adelaide—and the great event was finally recorded. The first female, carefully set up for photography, frustrated all hopes by vomiting six juveniles "at great speed, flying upwards . . . for approximately one meter . . . a substantial distance relative to the body size of the female." But the second mother obliged. Of the twenty-six offspring, two appeared gently and, apparently, voluntarily. The mother "partially emerged from the water, shook her head, opened her mouth, and two babies actively struggled out." The photo of a fully formed baby frog, resting on its parent's tongue before birth, has already become a classic of natural history. This second female, about two inches long, weighed 11.62 grams after birth. Her twenty-six children weighed 7.66 grams, or 66 percent of her weight without them. An admirable effort indeed!

Rheobatrachus inspired great excitement among Australian scientists, and research groups in Adelaide and Brisbane have been studying this frog intensively, with all work admirably summarized and discussed in a volume edited by M. J. Tyler (1983).[2] Rarely has such extensive and coordinated information been presented on a natural oddity, and we are grateful to these Australian scientists for bringing together their work in such a useful way.

This volume also presents enough detail (usually lacking in technical publications) to give nonscientists a feel for the actual procedures of research, warts and all (an appropriate metaphor for the subject). Glen Ingram's article on natural history, for example, enumerates all the day-to-day dilemmas that technical papers rarely mention: slippery bodies that elude capture; simple difficulties in seeing a small, shy frog that lives in inaccessible places (Ingram learned to identify *Rheobatrachus* by characteristic ripples made by its jump into water); rain, fog, and dampness; and regeneration that frustrates identification (ecologists must recognize individual animals in order to monitor size and movement of populations by mark-recapture techniques; amphibians and reptiles are traditionally marked by distinctive patterns of toe clipping, a painless and unobtrusive procedure, but *Rheobatrachus* frustrates tradition by regenerating its clipped toes, and Ingram could not reidentify his original captures). To this, we must add the usually unacknowledged bane of all natural history: boredom. You don't see your animals most of the time; so you wait and wait and wait (not always pleasant on the boggy banks of a stream in rainy season). Somehow, though, such plagues seem appropriate enough, given the subject. Frogs, after all, stand among the ten Mosaic originals: "I will smite all thy borders with frogs: And the river shall bring forth frogs abundantly, which shall go up and come into thine house, and into thy bedchambers, and upon thy bed . . . and into thine ovens, and into thy kneading troughs" (Exod. 8:2–3).

The biblical author of Exodus was, unfortunately, not describing *Rheobatrachus,* a rare animal indeed. Not a single *Rheobatrachus silus* has been seen in its natural habitat since 1981. A series of dry summers and late rains has restricted the range of this aquatic frog—and five years of no sightings must raise fears about extinction. Fortunately, a second species, named *R. vitellinus,* was discovered in January 1984, living in shallow sections of fast-flowing streams, about 500 miles north of the range of *R. silus.* This slightly larger version (up to three inches in length) also broods in its stomach; twenty-two baby frogs inhabited a pregnant female.

When discussed as a disembodied oddity (the problem with traditional writing in natural history), *Rheobatrachus* may pique our interest but not our intellect. Placed into a proper context among other objects of nature's diversity—the "comparative approach" so characteristic of evolutionary biology—gastric brooding in *Rheobatrachus* embodies a message of great theoretical interest. *Rheobatrachus,* in one sense, stands alone. No other vertebrate swallows its own fertilized eggs, converts its stomach into a brood pouch, and gives birth through its mouth. But in another sense, *Rheobatrachus* represents just one solution to a common problem among frogs.

In his review of parental care, R. W. McDiarmid argues that frogs display "the greatest array of reproductive modes found in any vertebrates" (see his article in G. M. Burghardt and M. Bekoff, 1978).[3] Much inconclusive speculation has been devoted to reasons for the frequent and independent evolution of brooding (and other forms of parental care) in frogs—a profound departure, after all, from the usual amphibian habit of laying eggs in water and permitting the young to pass their early lives as unattended aquatic tadpoles. Several authors have suggested the following common denominator: In many habitats, and for a variety of reasons, life as a free-swimming tadpole may become sufficiently uninviting to impose strong evolutionary pressure for bypassing this stage and undergoing "direct development" from egg to completed frog. Brooding is an excellent strategy for direct development—since tadpole life may be spent in a brood pouch, and the bad old world need not be faced directly before froghood.

In any case, brooding has evolved often in frogs, and in an astonishing variety of modes. As a minimal encumbrance and modification, some frogs simply attach eggs to their exteriors. Males of the midwife toad *Alytes obstetricans* wrap strings of eggs about their legs and carry them in tow.

At the other extreme of modification, some frogs have evolved special brood pouches in unconventional places. The female *Gastrotheca riobambae,* an Ecuadorean frog from Andean valleys, develops a pouch on her back, with an opening near the rear and an internal extension nearly to her head. The male places fertilized eggs in her pouch, where they develop under the skin of her back for five to six weeks before emerging as late-stage tadpoles.

In another Australian frog, *Assa darlingtoni,* males develop pouches on their undersides, opening near their hind legs but extending forward to the front legs (see article by G. J. Ingram, M. Anstis, and C. J. Corben, 1975).[4] Females lay their eggs among leaves. When they hatch, the male places himself in the middle of the mass and either coats himself with jelly from the spawn or, perhaps, secretes a slippery substance himself. The emerging tadpoles then perform a unique act of acrobatics among amphibians: they move in an ungainly fashion by bending their bodies, head toward tail, and then springing sideways and forward. In this inefficient manner, they migrate over the slippery body of their father and enter the brood pouch under their own steam. (I am almost tempted to say, given the Australian venue, that these creatures have been emboldened to perform in such unfroglike ways by watching too many surrounding marsupials, for the kangaroo's undeveloped, almost larval joey also must endure a slow and tortuous crawl to the parental pouch!)

Figure 9
Midwife toad *(Alytes obstetricans).* (From *The Riverside Natural History,* 1888, p. 329.)

In a kind of intermediate mode, some frogs brood their young internally but use structures already available for other purposes. I have already discussed *Rhinoderma,* the vocal-pouch brooder of Chile. Evolution seizes its opportunities. The male vocal pouch is roomy and available; in a context of strong pressure for brooding, some lineage will eventually overcome the behavioral obstacles and grasp this ready possibility. The eggs of *R. darwini* develop for twenty-three days before the tadpoles hatch. For the first twenty days, tadpoles grow within eggs exposed to the external environment. But tadpoles then begin to move, and this behavior apparently triggers a response from the male parent. He then takes the advanced eggs into his vocal pouch. They hatch there three days later and remain for fifty-two days until the end of metamorphosis, when the young emerge through their father's mouth as perfectly formed little froglets. In the related species *R. rufum,* muscular activity begins after eight days within the egg, and males keep the tadpoles in their vocal sacs for much shorter periods, finally expelling them, still in the tadpole stage, into water (see article by K. Busse, 1970).[5]

In this context, *Rheobatrachus* is less an oddity than a fulfillment. Stomachs provide the only other large internal pouch with an egress of sufficient size. Some lineage of frogs was bound to exploit this possibility. But stomachs present a special problem not faced by vocal sacs or novel pouches of special construction—and we now encounter the key dilemma that will bring us back to mimicry in butterflies and the evolutionary problem of incipient stages. Stomachs are already doing something else—and that something is profoundly inimical to the care and protection of fragile young. Stomachs secrete acid and digest food—and eggs and tadpoles are, as they say down under, mighty good tucker.

In short, to turn a stomach into a brood pouch, something must turn off the secretion of hydrochloric acid and suppress the passage of eggs into the intestine. At a minimum, the brooding mother cannot eat during the weeks that she carries young in her stomach. This inhibition may arise automatically and present no special problem. Stomachs contain "stretch receptors" that tell an organism when to stop eating by imposing a feeling of satiety as the mechanical consequence of a full stomach. A batch of swallowed eggs will surely set off this reaction and suppress further eating.

But this fact scarcely solves our problem—for why doesn't the mother simply secrete her usual acid, digest the eggs, and relieve her feeling of satiety? What turns off the secretion of hydrochloric acid and the passage of eggs into the intestine?

Tyler and his colleagues immediately realized, when they discovered gastric brooding in *Rheobatrachus,* that suppression of stomach function formed the crux of their problem. "Clearly," they wrote, "the intact amphibian stomach is likely to be an alien environment for brooding." They began by studying the changes induced by brooding in the architecture of the stomach. They found that the secretory mucosa (the lining that produces acid) regresses while the musculature strengthens, thus converting the stomach into a strong and chemically inert pouch. Moreover, these changes are not "preparatory"—that is, they do not occur before a female swallows her eggs. Probably, then, something in the eggs or tadpoles themselves acts to suppress their own destruction and make a congenial place of their new home. The Australian researchers then set out to find the substance that suppresses acid secretion in the stomach—and they have apparently succeeded.

P. O'Brien and D. Shearman, in a series of ingenious experiments, concentrated water that had been in contact with developing *Rheobatrachus* embryos to test for a chemical substance that might suppress stomach function in the mothers. They dissected out the gastric mucosa (secreting surface) of the toad *Bufo marinus* (*Rheobatrachus* itself is too rare to sacrifice so many adult females for such an experiment) and kept it alive in vitro. They showed that this isolated mucosa can function normally to secrete stomach acids and that well-known chemical inhibitors will suppress the secretion. They then demonstrated that water in contact with *Rheobatrachus* tadpoles suppresses the mucosa, while water in contact with tadpoles of other species has no effect. Finally, they succeeded in isolating a chemical suppressor from the water—prostaglandin E_2. (The prostaglandins are hormonelike substances, named for their first discovery as secretions of the human prostate gland—though they form throughout the body and serve many functions.)

Thus, we may finally return to mimicry and the problem of incipient stages. I trust that some readers have been bothered by an apparent dilemma of illogic and reversed causality. The eggs of *Rheobatrachus* must contain the prostaglandin that suppresses secretion of gastric acid and allows the stomach to serve as an inert brood pouch. It's nice to know that eggs contain a substance for their own protection in a hostile environment. But in a world of history—not of created perfection—how can such a system arise? The ancestors of *Rheobatrachus* must have been conventional frogs, laying eggs for external development. At some point, a female *Rheobatrachus* must have swallowed its fertilized eggs (presumably taking them for food, not with the foresight of evolutionary innovation)—and the fortuitous presence of prostaglandin suppressed digestion and permitted the eggs to develop in their mother's stomach.

The key word is *fortuitous*. One cannot seriously believe that ancestral eggs actively evolved prostaglandin because they knew that, millions of years in the future, a mother would swallow them and they would then need some inhibitor of gastric secretion. The eggs must have contained prostaglandin for another reason or for no particular reason at all (perhaps just as a metabolic by-product of development). Prostaglandin provided a lucky break with respect to the later evolution of gastric brooding—a historical precondition fortuitously available at the right moment, a sine qua non evolved for other reasons and pressed into service to initiate a new evolutionary direction.

Darwin proposed the same explanation for the initiation of mimicry—as a general solution to the old problem of incipient stages. Mimicry works splendidly as a completed system, but what gets the process started along one potential pathway among many? Darwin argued that a mimicking butterfly must begin with a slight *and fortuitous* resemblance to its model. Without this leg up for initiation, the process of improvement to mimetic perfection cannot begin. But once an accidental, initial resemblance provides some slight edge, natural selection can improve the fit from imperfect beginnings.

Thus, Darwin noted with pleasure Bates's demonstration that mimicry always arose among butterflies more prone to vary than others that never evolve mimetic forms. This tendency to vary must be the precondition that establishes fortuitous initial resemblance to models in some cases. "It is necessary to suppose," Darwin wrote, that ancestral mimics "accidentally resembled a member of another and protected group in a sufficient degree to afford some slight protection, this having given the basis for the subsequent acquisition of the most perfect resemblance." Ancestral mimics happened to resemble a model in some slight manner—and the evolutionary process could begin. The eggs of *Rheobatrachus* happened to contain a prostaglandin that inhibited gastric secretion—and their mother's stomach became a temporary home, not an engine of destruction.

New evolutionary directions must have such quirky beginnings based on the fortuitous presence of structures and possibilities evolved for other reasons. After all, in nature, as in human invention, one cannot prepare actively for the utterly unexpected. Gastric brooding must be an either-or, a quantum jump in evolutionary potential. As Tyler argues, what intermediary stage can one imagine? Many fishes (but no frogs) brood young in their mouths—while only males possess throat pouches, but only female *Rheobatrachus* broods in its stomach. Eggs can't develop halfway down the esophagus.

We glimpse in the story of *Rheobatrachus* a model for the introduction of creativity and new directions in evolution (not just a tale of growing bigger or smaller, fiercer or milder, by the everyday action of natural selection). Such new directions, as Darwin argued in resolving the problem of incipient stages, must be initiated by fortuitous prerequisites, thus imparting a quirky and unpredictable character to the history of life. These new directions may involve minimal changes at first—since the fortuitous prerequisites are already present, though not so utilized, in ancestors. A female *Rheobatrachus* swallowed its fertilized eggs, and a striking new behavior and mode of brooding arose at once by virtue of a chemical fortuitously present in eggs, and by the automatic action of stretch receptors in the stomach. Such minimal changes are pregnant with possibilities. Most probably lead nowhere beyond a few oddballs—as with *Rheobatrachus,* probably already well on its way to extinction.

But a few quirky new directions may become seeds of major innovations and floods of diversity in life's history. The first protoamphibian that crawled out of its pond has long been a favorite source of evolutionary cartoon humor. The captions are endless—from "see ya later as alligator" to "because the weather's better out here." But my favorite reads "here goes nothing." It doesn't happen often, but when nothing becomes something, the inherent power of evolution, normally an exquisitely conservative force, can break forth. Or, as Reginald Bunthorne proclaims in Gilbert and Sullivan's *Patience* (which evolution must have above all else): "Nature for restraint too mighty far, has burst the bonds of art—and here we are."

Postscript

It is my sad duty to report a change of state, between writing and republishing this essay, that has made its title eerily prophetic. *Rheobatrachus silus,* the stomach-brooding frog and star of this essay, has apparently become extinct. This species was discovered in 1973, living in fair abundance in a restricted region of southeast Queensland, Australia. In early 1990, the National Research Council (of the United States) convened a conference [in Irvine, California] to discuss "unexplained losses of amphibian populations around the world" (as reported in *Science News,* March 3, 1990). Michael J. Tyler, member of the team that discovered stomach brooding in *Rheobatrachus,* reported that 100 specimens could easily be observed per night when the population maintained fair abundance during the mid-1970s. Naturalists have not found a single individual since 1981, and must now conclude that the species is extinct (for several years they hoped that they were merely observing a sharp and perhaps cyclical reduction in numbers). Even more sadly, this loss forms part of a disturbing and unexplained pattern in amphibian populations throughout the world. In Australia alone, 20 of 194 frog species have suffered serious local drops in population size during the past decade, and at least one other species has become extinct.

DAVID SCOTT

"A Breeding Congress" (1998)

Had I given it much thought, I never would have expected to actually *hear* them. But there I was, standing in a quiet rain on a cool October night, listening to the marbled salamanders coming. Over leaves. Around logs. Tumbling off clumps of sedges. A few hundred had even crossed a nearby highway.

Sometimes while conducting field research on amphibians my hearing has played tricks on me. Was this my imagination? No. The marbled salamander herd thundered in as only salamanders can thunder, arriving by the thousands at a small, shrubby wetland in South Carolina called Ginger's Bay.

But first, what should we call this approaching horde? "Herd" is a bit misleading, with its connotation of trampling and mayhem. I prefer the term coined by a biologist decades ago: a breeding congress. Forget the trampling but keep the mayhem. And instead of lobbyist-filled hallways, picture low-lying depressions that fill with water part of the year—ephemeral wetlands called Carolina bays.

Marbled salamanders inhabit much of the eastern United States, but the coastal plain of the Carolinas—home to hundreds of these bays—is a hot spot. The marbled salamander's breeding cycle begins with a late summer or early autumn migration to the wetland site. Most of its close relatives (mole, tiger, and spotted salamanders) mate later, when the bays have filled with water, but marbled salamanders do their mating at the dry pond sites. After the animals breed, females deposit their eggs in nest cavities under vegetation and logs and in crayfish holes. They stay with the eggs for as long as a month. Once the ponds start to fill and the nests are flooded, the eggs hatch.

From there, marbled salamanders follow a relatively typical amphibian life cycle: aquatic larvae, a period of transition, then adult life on land. Typical, perhaps, but never dull. And fascinating enough that most of my rainy October nights during the last twelve years have been spent watching individuals of this one species, *Ambystoma opacum,* move from woodlands to wetlands on their annual breeding migration.

How the adults find the sites is a mystery—to us, not to them. At migration time, there is usually no water in the wetlands. The ponds fill later, with the onset of winter rains, so these pond-breeding salamanders are not cued by the presence of water. Males of the species have no mating calls (unlike male frogs and toads). Some scientists suspect that chemical signals are important because many individuals return to the pond in which they were born; perhaps they act as salmon do. Others suspect that the earth's magnetic field provides a cranial road map. I don't have a clue how they do it. Just call it radar love.

Marbled salamanders differ from many other salamanders, not only in their breeding time (early fall) and location (dry sites), but also in the predictability of their breeding efforts. Because they don't require the ponds to be filled, the timing of their migrations doesn't vary from year to year. Forget an explanation using scientific analyses and jargon. A few years ago, two British filmmakers asked to follow the migrations. With our assurances, of salamander regularity, they were able to book cheap airline seats several months in advance. We suggested they arrive on October 2, which they did. By 8:00 P.M. that night, the first pulse of several thousand males began to enter Ginger's Bay. There is no other species that I'd venture such a guarantee for, even with someone else's money at stake.

In many species of salamander, courtship activities are elaborate, with behavioral signals and mating dances that last for up to an hour before both individuals are "ready." At the peak of sexual stimulation, the male deposits a spermatophore—a packet of sperm on a jellylike base. At the appropriate stage of receptivity, the female salamander straddles a packet, squats, and incorporates the sperm mass into her cloaca. In describing this behavior, Lynne D. Houck, an ecologist at Oregon State University and a past president of the Society for the Study of Amphibians and Reptiles, actually managed to have the term "spermatophore play" published in a respected scientific journal. Such prolonged courtship, however, is not the case with marbled salamanders. The frenzied activity of the congress results in spermatophores being deposited everywhere; Houck called it a "minefield of spermatophores." Females hardly need to squat—they just pick up a spermatophore on the run. Such indiscriminate mating is a bit curious for a salamander species, although not unprecedented in the amphibian world. After all, a male southern toad "in the mood" will mate with your thumb.

On the right night, marbled salamander males are quite in the mood. From lab analyses of blood samples, we learned that these four-to-five-inch-long males have testosterone levels far exceeding mine. Males "dance" (their movements have been described as a waltz with a touch of slam dancing) with females and can deposit more than ten spermatophores in thirty minutes. Males also court other males, depositing spermatophores galore. Chin-rubbing with this animal, tail whipping with that one—gender doesn't seem to matter. What gives here? *A. opacum* has been around for far more than a million years. It must be doing something right. Last year at Ginger's Bay (which is about half the size of a football field), we caught almost 12,000 adults coming in to breed. Using mathematical models, we estimate the local population in the hardwood forest surrounding Ginger's Bay at close to 40,000 salamanders.

As you might expect, all the frenzy of searching for mates during the congress (which includes U-whipping tails, a male courtship behavior in which the animal jerks its tail to one side, temporarily changing its shape to that of the letter U) is energetically expensive. When male and female marbled salamanders enter the breeding site, they are fat and ready. Both sexes spend six to eight weeks at the wetland site without feeding. Many have close to 15 percent body fat, which appears, in females, to enable them to produce a clutch of 30–150 eggs, and in males, to fuel searching and courtship activities. A high proportion of the salamanders apparently don't survive; in some years the mortality during the breeding season is close to 50 percent. The reasons are not obvious. Skin toxins make adult marbled salamanders distasteful to many predators, so they aren't being eaten up. Some may die of old age—at eight, nine, or ten years, they've gone about as far as they can go. Adults that do survive the breeding season generally depart the wetland lean and hungry. Because future reproduction is contingent on accumulating adequate fat stores, some animals may take two or more years to prepare to breed again.

For newly hatched young, the food of choice is zooplankton—microscopic animals, such as daphnids and copepods, that live in the water. And there's not always enough to go around in the pond habitat. Many experiments have demonstrated that larval salamanders compete with one another for food. Jim Petranka, of the University of North Carolina at Asheville, manipulated numbers of larvae in entire ponds by adding or removing them (try catching several thousand slippery baby salamanders the size of shoelace tips). He found that more larvae mean smaller larvae. A higher density of babies in a pond results in slower growth and a longer development time (or larval period). These larvae emerge from the pond smaller, and that means fewer survive. I was amazed that salamanders that metamorphosed and left their natal pond at a small body size were still relatively small five or six years later, when they returned as prospective parents. Their larval environment had set a course for the rest of their ten-year lives.

I decided years ago that my personal goal in ecological research was to truly understand a single species well enough to make meaningful predictions. Just one species! At the Savannah River Ecology Laboratory, my colleagues and I have studied marbled salamanders from eggs to adulthood and beyond. In our experiments, we have manipulated females on their nests, numbers of larvae, aquatic predators (such as dragonflies and spiders), prey levels, timing of hatchings, water levels, and populations of coexisting amphibian species. We have studied salamander nest sites, salamander genetics, salamander metabolism, and salamander fat. At Rainbow Bay, another ephemeral wetland study site, we have collected data for twenty years on the natural population fluctuations of twenty-five amphibian species. It is the world's longest-running observational study of an entire amphibian community. The Rainbow Bay project provides a model for long-term ecological studies at a time when many amphibian species, particularly frogs, are thought to be in decline or in danger of disappearing. Using the data we gathered while observing natural populations in conjunction with our experiments, we have developed insights into the driving forces behind the commonly observed ups and downs of amphibian populations.

So you'd think I would know marbled salamanders by now. Yet I still owe my ex-office mate more pitchers of beer than I could ever repay due to errant predictions. Perhaps I'll get it right someday. It is encouraging to have learned so much, discouraging still to know so little, and disheartening that time and money are in short supply. Folks today (*my* Congress) attach too little importance to basic questions of ecology. If a critter is contaminated with mercury, exposed to radioactivity, or eliminated from vast stretches of habitat so that we have but a few left, then it may get some attention, especially if it is furry or pretty. As well it should. But what about understanding how to maintain "healthy" habitats and avoid costly fix-it plans in the first place? What about the value of understanding animal and plant populations that are relatively free from human disturbance?

Many questions remain, even for a species not in apparent decline. How do marbled salamanders find that wetland, anyway? What happens when it is destroyed, as is, sadly, often the case? We know that for this species a forested area around the wetland is essential, but we don't know how big this "buffer" strip must be. How "connected" must populations be if all are to survive? I hope I'll be able to help others answer these questions. But you'll know where to find me on a rainy October night: watching hormonally charged salamanders marching into Ginger's Bay. Once across the highway, they reach the safety of the wetland, and the tail-slapping mayhem begins. Finally, a congress I can relate to.

Part IV

Remembering the Earth
Twentieth-Century Nature Writers

With the cultural soil prepared by some influential eighteenth- and nineteenth-century literary pioneers, the Anglo-American nature writing tradition blossomed in the twentieth century. As a rising and increasingly urban human population has eaten further into wildland and the natural world in many areas has continued to deteriorate, the alienation of people and nature has become more acute. And absence, or distance, it seems, has in this case made many hearts grow fonder. The growing ranks of nature writers have prepared a lavish feast of organic fare for a reading public apparently hungry for some sense of connection with the nonhuman realm. Nature writers have avidly pursued and promoted this elemental remembering of the earth.

Modern science, of course, also directs one's gaze to the world of nature, but typically with a somewhat different quality of attention. Writer-naturalist Joseph Wood Krutch, echoing Henry David Thoreau and John Burroughs, drew the following distinction:

> Science is *knowledge about* natural phenomena while the proper subject of nature writing is an account of the writer's *experience with* the natural world. . . . He agrees that pure science must be unemotional and objective, but he doubts that even the scientist, much less the nonprofessional, can realize full human potentials if he is without wonder, or love, or a sense of beauty; if he never looks at the nature of which he is a part except with the cold eye of an outsider.[1]

Nature writers, on the whole, combine the Romantic's vision of the unity of nature, and human participation in this unity, with the scientist's eye for detail and systematic presentation of facts. Evolutionary theory has provided a valuable interpretative framework for a great many writers, and ecological science, with its emphasis on relationships, has also, especially since its ripening in the

1960s, afforded an abundance of welcome material. There is thus a definite, though sometimes subtle, radicalism to much nature writing because it often strikes at the root of the dominant dualism of human beings and nature that is part of our cultural inheritance.

The selections included in this part display a range of styles. Toward one end of the spectrum are works such as those by Edwin Way Teale and Ann Haymond Zwinger, in which careful description and communication of natural phenomena is the main feature and personal interpretation is a relatively minor element. Toward the other end are writings, George Orwell's and Joseph Wood Krutch's essays, for example, in which the events of nature are the impetus for a good deal of philosophical reflection. But the relative emphasis on information or interpretation is by no means clear-cut; all the pieces contain some of each, with the majority falling somewhere in the middle.

These writings also depict a variety of amphibian inhabitants of diverse locations. George Orwell takes pleasure in the vernal emergence of the common toad even in the middle of London. Dallas Lore Sharp is tutored by a tree toad in Massachusetts, whereas Donald Culross Peattie draws lessons from the wood frogs and spring peepers of Illinois. In Connecticut, the spring peepers also provide Joseph Wood Krutch with material for a sermon. Edwin Way Teale stops on a winter journey to examine long-tailed salamanders in Pennsylvania, and Annie Dillard reflects on the trials of life among frogs and newts in the salamander heaven of southern Appalachia. Northwestern salamanders and frogs afford Robert Michael Pyle many enjoyable forays into the dampness of evergreen western Washington State. In contrast, the dry redness of the Great Basin and the Grand Canyon are the terrain where Ann Haymond Zwinger and Terry Tempest Williams lovingly encounter spadefoot and red-spotted toads and the delightful spirits of frogs.

There is often a serious intent in much of this writing, with intimations of realities we are missing or admonitions about a world we have lost. But there is also typically a playfulness about it. It is evident that these writers relish the opportunity to ramble in landscapes from mundane to majestic and to communicate what they perceive and feel. Experiences in the natural world obviously evoke varied emotions, but one emotion that often surfaces is a simple joy at wandering loose in the wild—like a child, ready to waste time on a toad, disposed to wonder at a frog. Thomas Aquinas once said that "joy is the human's noblest act." Is the quiet joy of being at play among the creatures, at bottom, nature's creativity at play in us? Perhaps this exercise in enlarged, organic identity is, if truth be told, the nature writer's most scandalous act.

Dallas Lore Sharp
From *The Face of the Fields* (1911)
Chapter 3: "The Edge of Night"

The knoll yonder may be a kind of High Place, and its old apple tree a kind of altar for you when you had better not go to church, when your neighbor needs to be let alone, when your children are in danger of too much bread and of too many books—for the time when you are in need of that something which comes only out of the quiet of the fields at the close of day.

"But what is it?" you ask. "Give me its formula." I cannot. Yet you need it and will get it—something that cannot be had of the day, something that Matthew Arnold comes very near suggesting in his lines:—

> The evening comes, the fields are still.
> The tinkle of the thirsty rill,
> Unheard all day, ascends again;
> Deserted is the half-mown plain,
> Silent the swaths! the ringing wain,
> The mower's cry, the dog's alarms
> All housed within the sleeping farms!
> The business of the day is done,
> The last-left haymaker is gone.
> And from the thyme upon the height
> And from the elder-blossom white
> And pale dog-roses in the hedge,
> And from the mint-plant in the sedge,
> In puffs of balm the night-air blows
> The perfume which the day foregoes.

I would call it poetry, if it were poetry. And it is poetry, yet it is a great deal more. It is poetry and owls and sour apples and toads; for in this particular old apple dwells also a tree toad.

It is curious enough, as the summer dusk comes on, to see the round face of the owl in one hole, and out of another in the broken limb above, the flat weazened face of the tree toad. Philosophic countenances they are, masked with wisdom, both of them: shrewd and penetrating that of the slit-eyed owl; contemplative and soaring in its serene composure the countenance of the transcendental toad. Both creatures love the dusk; both have come forth to their open doors in order to watch the darkening; both will make off under the cover—one for mice and frogs over the meadow, the other for slugs and insects over the crooked, tangled limbs of the tree.

It is strange enough to see them together, but it is stranger still to think of them together, for it is just such prey as this little toad that the owl has gone over the meadow to catch.

Why does he not take the supper ready here on the shelf? There may be reasons that we, who do not eat tree toad, know nothing of; but I am inclined to believe that the owl has never seen his fellow lodger in the doorway above, though he must often have heard him piping his gentle melancholy in the gloaming, when his skin cries for rain!

Small wonder if they have never met! for this gray, squat, disc-toed little monster in the hole, or flattened on the bark of the tree like a patch of lichen, may well be one of those things which are hidden from the sharp-eyed owl. Whatever purpose be attributed to his peculiar shape and color,—protective, obliterative, mimicking,—it is always a source of fresh amazement, the way this largest of our hylas, on the moss-marked rind of an old tree, can utterly blot himself out before your staring eyes.

The common toads and all the frogs have enemies enough, and it would seem from the comparative scarcity of the tree toads that they must have enemies, too, but I do not know who they are. The scarcity of the tree toads is something of a puzzle, and all the more to me, that, to my certain knowledge, this toad has lived in the old Baldwin tree, now, for five years. Perhaps he has been several, and not one; for who can tell one tree toad from another? Nobody; and for that reason I made, some time ago, a simple experiment, in order to see how long a tree toad might live, unprotected, in his own natural environment.

Upon moving into this house, about seven years ago, we found a tree toad living in the big hickory by the porch. For the next three springs he reappeared, and all summer long we would find him, now on the tree, now on the porch, often on the railing and backed tight up against a post. Was he one or many? we asked. Then we marked him; and for the next four years we knew that he was himself alone. How many more years he might have lived in the hickory for us all to pet, I should like to know; but last summer, to our great sorrow, the gypsy-moth killers, poking in the hole, did our little friend to death.

He was worth many worms.

It was interesting, it was very wonderful to me, the instinct for home—the love for home I should like to call it—that this humble little creature showed. A toad is an amphibian to the zoologist; an ugly gnome with a jeweled eye to the poet; but to the naturalist, the lover of life for its own sake, who lives next door to his toad, who feeds him a fly or a fat grub now and then, who tickles him to sleep with a rose leaf, who waits as thirstily as the hilltop for him to call the summer rain, who knows his going to sleep for the winter, his waking up for the spring—to such an one the jeweled eye and the amphibious habits are but the forewords of a long, marvelous life-history.

This small tree toad had a home, had it in his soul, I believe, precisely where John Howard Payne had it, and where many another of us has it. He had it in a tree, too,—in a hickory tree, this one that dwelt by my house; he had it in an apple tree, that one yonder across the meadow.

East, west,

Hame's best,

croaked our tree toad in a tremulous, plaintive minor that wakened memories in the vague twilight of more old, unhappy, far-off things than any other voice I ever knew.

These two tree toads could not have been induced to trade houses, the hickory for the apple, because a house to a toad means home, and a home is never in the market. There are many more houses in the land than homes. Most of us are only real-estate dealers. Many of us have never had a home; and none of us has ever had more than one. There can be but one—mine—and that has always been, must always be, as imperishable as memory, and as far beyond all barter as the gates of the sunset are beyond my horizon's picket fence of pines.

The toad seems to feel it all, but feels it whole, not analyzed and itemized as a memory. Here in the hickory for four years (for seven, I am quite sure) he lived, single and alone. He would go down to the meadow when the toads gathered there to lay their eggs, but back he would come, without mate or companion, to his tree. Stronger than love of kind, than love of mate, constant and dominant in his slow cold heart is his instinct for home.

If I go down to the orchard and bring up from the apple tree another toad to dwell in the hole of the hickory, I shall fail. He might remain for the day, but not throughout the night, for with the gathering twilight there steals upon him an irresistible longing, the *Heimweh* which he shares with me; and guided by it, as the bee and the pigeon and the dog are guided, he makes his sure way back to the orchard home.

Would he go back beyond the orchard, over the road, over the wide meadow, over to the Baldwin tree, half a mile away, if I brought him from there? We shall see. During the coming summer I shall mark him in some manner, and bringing him here to the hickory, I shall then watch the old apple tree yonder. It will be a hard, perilous journey. But his longing will not let him rest; and guided by his mysterious sense of direction—for this *one* place—he will arrive, I am sure, or he will die on the way.

Yet I could wish there were another tree here, besides the apple, and another toad. Suppose he never gets back? Only one toad less? A great deal more than that. Here in the old Baldwin he has made his home for I don't know how long, hunting over its world of branches in the summer, sleeping down in its deep holes during the winter—down under the chips and punk and castings, beneath the nest of the owls, it may be; for my toad in the hickory always buried himself so, down in the debris at the bottom of the hole, where, in a kind of cold storage, he preserved himself until thawed out by the spring. I never pass the old apple in the summer but that I stop to pay my respects to the toad; nor in the winter that I do not pause and think of him asleep in there. He is no mere toad any more. He has passed into a *genius loci,* the Guardian Spirit of the tree, warring in the green leaf against worm and grub and slug, and in the dry leaf hiding himself, a heart of life, within the tree's thin ribs, as if to save the old shell to another summer.

A toad is a toad, and if he never got back to the tree there would be one toad less, nothing more. If anything more, then it is on paper, and it is cant, not toad at all. And so, I suppose, stones are stones, trees trees, brooks brooks—not books and tongues and sermons at all—except on paper and as cant. Surely there are many things in writing that never had any other, any real existence, especially in writing that deals with the out-of-doors. One should write carefully about one's toad; fearfully, indeed, when that toad becomes one's teacher; for teacher my toad in the old Baldwin has many a time been.

Often in the summer dusk I have gone over to sit at his feet and learn some of the things my college professors could not teach me. I have not yet taken my higher degrees. I was graduated A. B. from college. It is A. B. C. that I am working toward here at the old apple tree with the toad.

Seating myself comfortably at the foot of the tree, I wait; the toad comes forth to the edge of his hole above me, settles himself comfortably, and waits. And the lesson begins. The quiet of the summer evening steals out with the wood-shadows and softly covers the fields. We do not stir. An hour passes. We do not stir. Not to stir is the lesson—one of the majors in this graduate course with the toad.

The dusk thickens. The grasshoppers begin to strum; the owl slips out and drifts away; a whippoorwill drops on the bare knoll near me, clucks and shouts and shouts again, his rapid repetition a thousand times repeated by the voices that call to one another down the long empty aisles of the swamp; a big moth whirs about my head and is gone; a bat flits squeaking past; a firefly blazes, but is blotted out by the darkness, only to blaze again, and again be blotted, and so passes, his tiny lantern flashing into a night that seems the darker for the quick, unsteady glow.

We do not stir. It is a hard lesson. By all my other teachers I had been taught every manner of stirring, and this unwonted exercise of being still takes me where my body is weakest, and it puts me painfully out of breath in my soul. "Wisdom is the principal thing," my other teachers would repeat, "therefore get wisdom, but keep exceedingly busy all the time. Step lively. Life is short. There are *only* twenty-four hours to the day. The Devil finds mischief for idle hands to do. Let us then be up and doing"—all of this at random from one of their lectures on "The Simple Life, or the Pace that Kills."

Of course there is more or less of truth in this teaching of theirs. A little leisure has no doubt become a dangerous thing—unless one spends it talking or golfing or automobiling, or aëroplaning or elephant-killing, or in some other diverting manner; otherwise one's nerves, like pulled candy, might set and cease to quiver; or one might even have time to think.

"Keep going,"—I quote from another of their lectures,—"keep going; it is the only certainty you have against knowing whither you are going." I learned that lesson well. See me go—with half a breakfast and the whole morning paper; with less of lunch and the 4:30 edition. But I balance my books, snatch the evening edition, catch my car, get into my clothes, rush out to dinner, and spend the evening lecturing or being lectured to. I do everything but think.

But suppose I did think? It could only disturb me—my politics, or ethics, or religion. I had better let the editors and professors and preachers think for me. The editorial office is such a quiet thought-inducing place; as quiet as a boiler factory; and the thinkers there, from editor-in-chief to the printer's devil, are so thoughtful for the size of the circulation! And the college professors, they have the time and the cloistered quiet needed. But they have pitiful salaries, and enormous needs, and their social status to worry over, and themes to correct, and a fragmentary year to contend with, and Europe to see every summer, and—Is it right to ask them, with all this, to think? We will ask the preachers instead. They are set apart among the divine and eternal things; they are dedicated to thought; they have covenanted with their creeds to think; it is their business to study, but, "to study to be careful and harmless."

It may be, after all, that my politics and ethics and religion need disturbing, as the soil about my fruit trees needs it. Is it the tree? or is it the soil that I am trying to grow? Is it I, or my politics, my ethics, my religion? I will go over to the toad, no matter the cost. I will sit at his feet, where time is nothing, and the worry of work even less. He has all time and no task; he is not obliged to labor for a living, much less to think. My other teachers all are; they are all professional thinkers; their thoughts are words: editorials, lectures, sermons,—livings. I read them or listen to them. The toad sits out the hour silent, thinking, but I know not what, nor need to know. To think God's thoughts after Him is not so high as to think my own after myself. Why then ask his of the toad, and so interrupt these of mine? Instead we will sit in silence and watch Altair burn along the shore of the sky, and overhead Arcturus, and the rival fireflies flickering through the leaves of the apple tree.

The darkness has come. The toad is scarcely a blur between me and the stars. It is a long look from him, ten feet above me, on past the fireflies to Arcturus and the regal splendors of the Northern Crown—as deep and as far a look as the night can give, and as only the night can give. Against the distant stars, these ten feet between me and the toad shrink quite away; and against the light far off yonder near the pole, the firefly's little lamp becomes a brave but a very lesser beacon.

There are only twenty-four hours to the day—to the day and the night! And how few are left to that quiet time between the light and the dark! Ours is a hurried twilight. We quit work to sleep; we wake up to work again. We measure the day by a clock; we measure the night by an alarm clock. Life is all ticked off. We are murdered by the second. What we need is a day and a night with wider margins—a dawn that comes more slowly, and a longer lingering twilight. Life has too little selvage; it is too often raw and raveled. Room and quiet and verge are what we want, not more dials for time, nor more figures for the dials. We have things enough, too, more than enough; it is space for the things, perspective, and the right measure for the things that we lack—a measure not one foot short of the distance between us and the stars.

If we get anything out of the fields worth while, it will be this measure, this largeness, and quiet. It may be only an owl or a tree toad that we go forth to see, but how much more we find—things we cannot hear by day, things long, long forgotten, things we never thought or dreamed before.

The day is none too short, the night none too long; but all too narrow is the edge between.

From *Sanctuary! Sanctuary!* (1926)
Chapter 3: "My Twenty-Four-Dollar Toad"

Probably my toad would not fetch twenty-four dollars on the auction block. I certainly did not pay twenty-four dollars for him; nor am I advertising him for sale at that price. He is not mine to buy and sell, being mine only because he lives in my stone wall by the steps. If he belongs to anything it is to the place—the wall, the yard, the garden, and the sky and the stars, his companions through the night.

He is not so free as the winds to come and go, but he is as free as his slow squat legs can make him. This price is put upon him by the biological department of the government, as a measure of his value. This is a fair price, and an easy sort of arithmetic. There are men who can count in dollars and in no other terms.

As for me, I don't like the dollar-and-cent talk about toads and birds. There are other ways of valuing things than by the dollar mark. With my toad, however, it is interesting to have his practical value figured out exactly. I have always known him to be a useful citizen, but now that I can be taxed, possibly, for every toad in my garden, I certainly am going to guard him jealously.

But I should like to do a little hop-toad figuring with a pencil of my own. The man who figured this twenty-four-dollar value out didn't have a garden. That I am sure. He arrived at this figure with the help of reports and statistics. Those things cannot tell the truth, at least not the whole truth.

In my garden, among many other things, I have eighteen hills of watermelons, three vines to the hill. One cutworm, in one night, could destroy the three vines in any hill. But a big brown toad goes up and down and over and across that melon-patch every night catching cutworms. Suppose each night he catches the cutworm that had plotted to eat off the three vines in one of the melon hills. How much would I owe that toad at the end of the watermelon season?

Let me work that problem out, for no one else can do it, because no one else knows how much a watermelon costs me. I can buy one at the store for a dollar. But that's a store melon. You can't compare a store melon with a home-grown melon. The home-grown melon is very much higher in quality. It is also very much higher in price. I don't believe anybody knows exactly the cost price of a home-grown watermelon in Massachusetts. Perhaps twenty-four dollars. Perhaps more.

Say it is an average of twenty-four dollars each that a watermelon costs in my garden, and that each of the three vines in each of the eighteen hills, thanks to the watchful toad, bears one melon. That makes fifty-four melons, which, at twenty-four dollars each, come to $1,296.00—the tidy little sum I owe at the end of the watermelon season to my toad.

Now there may be an error somewhere in these figures. I may have selling price and cost price mixed up here. Certainly I should hate to sell one of my home-grown watermelons for twenty-four dollars. They are beyond price, so much labor, and love, and waiting, and disappointment enter into them; and often so little juice and sugar and core that I couldn't give one of them away, to say nothing of selling it.

These sums, I say, may not be wholly correct. Something must be allowed for what the scientists call *the personal equation*. A scientific toad may be worth precisely what the scientists say he is. But my toad is a personal toad—that's a different toad altogether, and a different price. And so with melons, and a whole world of things.

My farmer brother in New Jersey was at work in his sweet-potato patch lately when he accidentally got mixed up with a toad and a very doubtful affair.

The rows of sweet-potato hills were highly ridged, and as he was hoeing the young vines he noticed a large hog-nosed adder snake coming up the valley between two of the rows toward him. The snake was plainly looking for something. Its head was slightly raised, and if it had been a dog, he would have said it was sniffing the wind for some scent.

Do snakes hunt by scent? That is a question I should like to have answered, for surely this snake seemed to be so hunting.

Swinging forward with the sinuous motion of the gliding body, the head of the creature moved about on the level of the ridged rows, now along one row, then over to the other, when suddenly it rose a little higher and hung motionless and poised.

The snake was surely hunting. But was it now listening, or had it caught the *scent* of its prey? And what prey?

Man and snake were face to face and less than the length of a hoe apart; but the yellow eyes of the adder saw nothing—nothing more than a shadow, nothing for alarm. Something nearer to it than the man held its half-sprung body stiff in the air with attention.

Slowly gathering its length together, the snake, satisfied that it was right, deliberately thrust its shovel-like nose into the sand under one of the potato hills, until its head was buried to the neck, when, *flop!* and out on the opposite side of the hill tumbled a big toad who made off down the furrow as if Death were after him.

And Death was after him. Instantly the adder pulled out his head, went over the ridge, and, apparently picking up the scent, started swiftly on the hot trail.

The toad had lost no time getting off his mark, and, before the snake was well under way behind him, had taken a good twenty hops, climbed the steep ridge of hills, and dropped out of sight in the furrow on the other side.

Nor had the snake yet seen him. And this is what makes it seem the certain work of the reptile's nose, for without the slightest hesitation the adder, like an awful fate, was in pursuit, going back down this row. Coming to the point where the toad had climbed the ridge, the snake climbed over also, and into the furrow, reversed himself, and now came once more up the row toward my brother, the toad in sight and losing ground at every hop.

The man through all the little tragedy had not stirred a muscle. On came the frightened toad, going as probably the short, dumpy thing had never gone before, straight at my brother. It landed by his feet, took another feeble hop, and dived between them, just as the hungry adder got within striking distance of the long hoe.

Few men in the face of a snake think twice before they act. There was a flash of steel in the air and the chase was over. The adder literally bit the dust, as its habit is when hurt. The toad hopped on in safety, while my brother went about his hoeing, twenty-four dollars saved, if the Biological Survey is correct, but very much out of pocket as an observer, and still more as a philosopher, for his hasty act, wondering what right he had to kill the snake, and if he had not made matters worse instead of better in his small world of reptiles and toads and sweet-potatoes.

I have a feeling that he did make matters very much worse. It is just a feeling. The toad is doubtless worth twenty-four dollars. I do not know how much the snake is worth. But I have a feeling that he, too, is worth twenty-three or twenty-four dollars, though I cannot "prove" it as I used to my answers in arithmetic. This is a kind of "profit and loss" problem. Those in the arithmetic were hard enough. This one is harder than any there.

For who knows enough to say: "This toad will bring me profit. This snake will cause me loss. So I will spare the toad and reap the profit; but I will scotch the snake and avoid the loss"? A man's world is bigger than his sweet-potato patch. Most things have many values, but there are few things so highly priced as life. A dead snake is worthless truly. Can this be possible also of a snake's life?

Yesterday I bent down a walnut sapling along a shady street in order to peek into a red-eyed vireo's nest. Behold, one vireo egg and three eggs of the cowbird! I took out the three evil cowbird eggs and told the young boy who was with me to destroy them.

I saved one good vireo at the cost of three bad cowbirds. Who says "good" and "bad"? Today in my neighbor's meadow I watched a little band of cowbirds feeding about the noses and under the very feet of the cattle, hopping up and settling close to the Ayrshires as they grazed slowly along the stream. If cattle could observe and write, what would they say of good vireos and bad cowbirds?

On the road to the village is a well-kept garden, open to the rain, sloping toward the south, of warm, sweet, mellow soil in perfect tilth, a lovely, living thing, the joint work of human hands and hands within the earth and from the sky. In the middle of the garden, like a felon from a gibbet, swings the mummied body of a crow.

It is a black thing—blacker than its raven feathers; stake and string and dried distorted wings the sole work of human hands. The bird had come to the garden in the spirit of the sunshine and the rain, a helper straight from nature to get the beetle and the grub. But he had also helped himself to the sprouting corn.

Hang him head down in the middle of the garden! Publish to every passing crow that this shall be his fate, also, should he venture under the open sky within gunshot of the garden! And publish to every passing man along the street that this is a human garden, a garden with a gun as well as a hoe.

However good the corn crop in that garden, there is something evil growing there. A hedge of hollyhocks and old-fashioned flowers incloses the garden, but the black thing on the gibbet darkens their bright faces and leaves all the glory stained. The blessing of Heaven is on the hoe, but on the gun is the curse of Cain.

There must be a better way. After planting my corn I hang a wide shingle from the top of a slanting stake in the garden, the string run through the *thick* end, the thin end free to spin and toss in every breeze. The crows never pull my sprouting corn. Except for stealing a few sweet cherries, which chance to ripen just as the young crows get a-wing, the crows from corn-planting time to melon time do only good to my garden, and no suspicious shingle turns in the wind, once the corn is well up, to warn them away.

It is hard to walk upon this green earth without stepping on the grass. There is so much grass. So is it hard to plant a little garden in the wilderness and keep the wilderness out. Cat-like, round and round the thatched village creeps the jungle, ever trying to get in. We invented the gun to help stand it off, but the older, simpler, more effective tool is the hoe. Heaven can be brought to earth by hoeing; it is desolation that follows the gun.

Nevertheless, the crow and the cowbird and the snake are problems, while it seems quite certain that I have a twenty-four-dollar toad. The other night as the car climbed Mullein Hill and the search-lights swept the dooryard and flashed into the open barn, there on the barn-door sill sat my toad, directly in the path of the wheels, staring blindly into the glaring lights.

Pulling up short, I waited for him to hop off. One is not going to run over a twenty-four-dollar toad if one can help it. A little knowledge, an old adage says, is a dangerous thing, but not if it is as much as twenty-four dollars' worth. That one bit of knowledge got hold of by the public about the toad would greatly increase the lowly batrachian's margin of safety. If the public could get hold of still more knowledge about the toad, about his soulful singing in the spring, his strange homing instinct, his extraordinary habits—drinking through his skin, eating up his old clothes—the public would find him even more interesting than he is profitable, and so make his humble place in the dooryard and garden both beautiful and secure.

"Ugh!" exclaimed a very *nice* little girl, "he is so bunchy and squashy and big-mouthed and homely!" So he is. He is not a bit like a butterfly. Nobody would mistake him in the garden for a pink or a pansy. Call him ugly if you dare to, but don't let the fairies hear you. For the toad is a fairy, if you only knew. If not a fairy, he is a goblin, a good goblin, and only missed being a fairy by a margin of wings. He has lovely bow legs instead—on the front; and a horizontal pair behind.

Goblins gobble and fairies are fair, and there is a great difference in that. Let evening come, and the day's work be finished, and the soft twilight creep down over the garden, then who shall guard the tender lettuces and the sprouting beans? We house-folk will soon be fast asleep.

With the dusk comes the dew, and together they creep beneath a little melon-vine and whisper: "Wake up, sleepyhead, and wink the dirt out of your eyes. The hoe-man has gone. The cutworms are crawling, and a bushel of bugs are marching on the garden. Wake up!"

Then the earth heaves, cracks, and falls away, uncovering a big pop-eye which blinks laboriously once at the damp and once at the gloom. This exertion leaves everything at a standstill. The wink takes place again. Then the standstill takes place for another extended time. But now the soil is bulging, breaking, and, as you live, a clod of earth with eyes and legs is emerging! Or is it a troll?

A troll it is, if ever there was a troll, a dwarf, droll troll, squat and humped and warty. And off he hops with the earth of his bed still sticking to his back. Then he stops to look and listen! But not to look at the fireflies sputtering over the garden, nor to listen to the whippoorwills whimpering in the wood lot.

Something is moving in the lettuce row! Something small and still, so still that a hoe-man's ears could never hear it. The goblin troll turns his head, stretches his baggy body till it is taut, that he may the better hear. Then he turns the rest of himself around, takes a step, a hop, a jump—and flash! Out shoots a long tongue and in it shoots, and the cutworm shoots in with it. And the less we say about that cutworm, the more we can say about the lettuce.

Up and down the lettuce row, up and down the bean row, up and down the onion row, up and down the radish row, up and down the tomato row, up and down and over and across the rows and rows hops the good goblin, missing more grubs and bugs than he finds, I am sure, but finding enough to keep the garden safe till daylight comes, and the hoe-man comes to carry on.

In the beet row a dozen baby plants lie withered on their sides. "A cutworm!" growls the hoe-man. A little squash-vine is wilted. "The striped beetle!" fumes the hoe-man. A stalk of tender corn looks curled and sickly yellow. "The horrid borer!" swears the hoe-man, scratching the ground like a chicken, his foot within an inch of a melon-vine under whose shelter, buried out of sight in the soil, sleeps the good goblin.

The goblin's round little belly is bulging with cutworms and striped beetles and borers. The goblin is very comfortable and, like *Joe, the fat boy,* is very sound asleep. If the man with the hoe could count all the ingredients of the goblin's comfort, and the numbers, then reckon the numbers of beets and onions and melons and lettuce still standing in their rows, until he had the exact equivalent of the goblin's contents, and could reckon their value in terms of the green grocer, he might find the good goblin was sleeping off about twenty-four dollars' worth of vegetables, taken as concentrated insects in a single night!

That, at least, is the way I figure it, being a gardener.

Donald Culross Peattie
From *An Almanac for Moderns* (1935)

March Eleventh. I have said that much of life and perhaps the best of it is not quite "nice." The business of early spring is not; it transpires in nakedness and candor, under high empty skies. Almost all the first buds to break their bonds send forth not leaves but frank catkins, or in the maple sheer pistil and stamen, devoid of the frilled trimmings of petals. The cedar sows the wind with its pollen now, because it is a relict of an age before bees, and it blooms in a month essentially barren of winged pollinators. The wood frogs, warmed like the spring flowers by the swift-heating earth, return to the primordial element of water for their spawning, and up from the oozy bottoms rise the pond frogs, to make of the half-world of the marges one breeding ground.

It is a fact that the philosopher afoot must not forget, that the astonishing embrace of the frog-kind, all in the eery green chill of earliest March, may be the attitude into which the tender passion throws these batrachians, but it is a world away from warm-blooded mating. It is a phlegmatic and persisting clasping, nothing more. It appears to be merely a reminder to the female that death brings up the rear of life's procession. When after patient hours he quits her, the female goes to the water to pour out her still unfertilized eggs. Only then are they baptized with the fecundating complement of the mate.

It is a startling bit of intelligence for the moralists, but the fact seems to be that sex is a force not necessarily concerned with reproduction; back in the primitive one-celled animals there are individuals that fuse without reproducing in consequence; the reproduction in those lowly states is but a simple fission of the cell, a self-division. It seems then that reproduction has, as it were, fastened itself on quite another force in the world; it has stolen a ride upon sex, which is a principle in its own right.

March Twelfth. Today winter has returned in a tantrum. The ponds, turned bitter gray, look as if ice would gather on them again, and in the high leafless hardwood groves the wind flings about and stamps on the trembling shoots of wildflowers; it takes the wood in a fury, setting up a great roaring upon a single tone, high overhead, as if it had found the keynote of the trees and were vibrating them to the root. So, house-bound and angry for it, I add a few words less pleasing, perhaps, to the moralists than what has gone before.

A long survey of the ascent of sex has shown all who ever made it that the purpose of this awe-inspiring impulsion is nothing more, nor less, than the enrichment of life. Reproduction purely considered gets on a great deal better without anything so chancy as mating. What sex contributes to it is the precious gift of variation, as a result of commingling. And as variety is the spice of life, it has come to be—thanks to the invitation of sex which creatures accept with such eagerness—one of life's chief characteristics. Thus sex is what the lover has always wished to believe, a worthy end in itself. It is to be revered for its own sake, and the very batrachians know it.

March Thirteenth. It is a complaint of the poets that men of science concern themselves not at all with beauty. But the scientists mind their own business, and they know that all men mean something different by beauty. Rodin preferred men with broken noses, old hags of the street; Romney liked handsome high-born children. To speak broadly, the variety of life is its beauty; you may choose out of that as you will.

In the age of piety, it was supposed that the purpose of living beauty was to be useful. No less a man than Darwin proposed the idea of sexual selection. The breast of a grosbeak was colored to win a mate, the catbird sang in competition with his fellows to win the little female away from them, and thus the whole duty of birds—to be fruitful and multiply—was advanced. You may pretend so if you like, but it is not demonstrable. It is even to be doubted that the color of the flowers serves any such righteous purpose as attracting the bee; many flower-haunting insects seem blind to color. The beauty of a butterfly's wing, the beauty of all things, is not a slave to purpose, a drudge sold to futurity. It is excrescence, superabundance, random ebullience, and sheer delightful waste to be enjoyed in its own high right.

March Fourteenth. I set forth on this high-promising day for the hills, but the slope of the land drew me downward; the brook, running toward the river, led me on, and I walked as a man who knows that the day has something in store for him, something it would disclose. Before very long I was at the lowest level in this neighborhood, on the springy turf, full of green spears of coarse grass, of the river meadows. The sunlight hung in the misty willows. *Pee-yeep . . .*

pee-yeep came the sweet metallic clink of the spring peepers, but when I tried to stalk them, ever so quietly, I was forestalled, surrounded by silence, a man alone on the wild useless bottom land, under the remote candor of the skies that arched the marsh and me. *Pee-yeep*—like the horizon, the sweet melancholy sound receded or closed up behind me.

Among a penciling of last year's reeds, upon the very marge, I stood and saw the frog's eggs in the water. Laid only today, perhaps, the dark velvety globules in their sphere of silver jelly shine softly up at me, reminding me that a year ago I was seeing them in this same pond, and proposing to have their secret. Well, the Ides of March are here, they say. Can I recite the ritual of their ancient freemasonry?

March Fifteenth. I lift from the chill and cloudy waters of the woodland marsh a bit of frog's egg jelly, and the very feel of it on my fingers is dubious and suggestive. I accept without even cortical repulsion the sensation of their mucous envelope, for I have grown used by now to the gelatinous feel that conveys the very nature of protoplasm. This plasmic feel, traveling up from my nerve ends, asks questions now of my brain.

Suppose that these eggs are so fresh-laid that they are still unfecundated; are they yet come to life? At what instant does individual life begin? Usually, we deem, from the moment that sperm meets egg. Before that happens are the ova half alive? Or do they not simply represent pure potentiality, such as Aristotle meant when he called the rock in the quarry, awaiting the sculptor's chisel, potentiality?

It is just here that the mechanistic biologists have sought to drive in their wedge. They removed the frog's eggs before they were fecundated, and essayed to stimulate them "into life" without benefit of fatherhood. And they discovered that chemicals, or even a mere pinprick in the nucleus, would start the unfertilized egg cell to dividing and developing. In a few brief weeks the half-orphan tadpoles were grown to clamorous, croaking frogs!

March Sixteenth. Almost it seems as if the great mechanist, Jacques Loeb, and the clever laboratory man, Bataillon, had found the break in the charmed circle of life.

But the vitalists are ready, with the well taken reply that because a chemical or a mechanical irritation will stimulate the unfertilized frog eggs to develop parthenogenetically—that is, by immaculate conception—it does not follow that conception is but a physico-chemical process as Loeb has boldly stated. For in their preoccupations with acids and needle pricks, the mechanists have forgotten to examine the most remarkable feature of all—the nature of the egg itself. All that the needle and the reagent did was to release the forces of cleavage, development and metabolism that were already stored within the egg. The closed, charmed circle of life remains intact, was never broken.

Behold Driesch grinding the eggs of Loeb's favorite sea urchin up between plates of glass, pounding and breaking and deforming them in every way. And when he ceased from thus abusing them, they proceeded with their orderly and normal development. Is any machine conceivable, Driesch asks, which could thus be torn down into parts and have each part continue to act like a whole machine? Could any machine have its parts all disarranged and transposed, and still have them act normally? One cannot imagine it. But of the living egg, fertilized or not, we can say that there lie latent within it all the potentialities presumed by Aristotle, and all of the sculptor's dream of form, yes, and the very power in the sculptor's arm.

March Seventeenth. So at the end of it all we come to a truce between the old wrestlings of a mechanistic with a vitalistic view of life. We grant, and gladly, with a sense of kinship to the great elements and forces, that life is built of the same star-stuff as the rest of the universe. It obeys physical and chemical laws. But I am of those who believe that it is a law in itself.

A thing is either alive or it isn't; there is nothing that is almost alive. There is but the remotest possibility of the origin of life by spontaneous generation, and every likelihood that Arrhenius is right when he dares to claim that life is a cosmic phenomenon, something that drifts between the spheres, like light, and like light transiently descends upon those fit to receive it. Life is a phenomenon, *sui generis,* a primal fact in its own right, like energy. Cut flesh or wood how you like, hack at them in a baffled fury—you cannot find life itself, you can only see what it built out of the lifeless dust. Can you see energy in a cresting wave, a shaft of spring sunlight? No, energy is but a name for something absolutely primal which we cannot analyze or comprehend but only measure in science and depict in art. Life, too, is an ineffable, like thought. It is the glory on the earth.

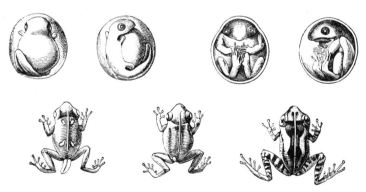

Figure 10
Development of Puerto Rican coqui *(Eleutherodactylus coqui).*
(From *The Riverside Natural History,* 1888, p. 331.)

George Orwell

"Thoughts on the Common Toad" (1946)

Before the swallow, before the daffodil, and not much later than the snowdrop, the common toad salutes the coming of spring after his own fashion, which is to emerge from a hole in the ground, where he has lain buried since the previous autumn, and crawl as rapidly as possible towards the nearest suitable patch of water. Something—some kind of shudder in the earth, or perhaps merely a rise of a few degrees in the temperature—has told him that it is time to wake up: though a few toads appear to sleep the clock round and miss out a year from time to time—at any rate, I have more than once dug them up, alive and apparently well, in the middle of the summer.

At this period, after his long fast, the toad has a very spiritual look, like a strict Anglo-Catholic toward the end of Lent. His movements are languid but purposeful, his body is shrunken, and by contrast his eyes look abnormally large. This allows one to notice, what one might not at another time, that a toad has about the most beautiful eye of any living creature. It is like gold, or more exactly it is like the golden-colored semi-precious stone which one sometimes sees in signet rings, and which I think is called a chrysoberyl.

For a few days after getting into the water the toad concentrates on building up his strength by eating small insects. Presently he has swollen to his normal size again, and then he goes through a phase of intense sexiness. All he knows, at least if he is a male toad, is that he wants to get his arms round something, and if you offer him a stick, or even your finger, he will cling to it with surprising strength and take a long time to discover that it is not a female toad. Frequently one comes upon shapeless masses of ten or twenty toads rolling over and over in the water, one clinging to another without distinction of sex. By degrees, however, they sort themselves out into couples, with the male duly sitting on the female's back. You can now distinguish males from females, because the male is smaller, darker and sits on top, with his arms tightly clasped round the female's neck. After a day or two the spawn is laid in long strings which wind themselves in and out of the reeds and soon become invisible. A few more weeks, and the water is alive with masses of tiny tadpoles which rapidly grow larger, sprout hind legs, then forelegs, then shed their tails: and finally, about the middle of the summer, the new generation of toads, smaller than one's thumbnail but perfect in every particular, crawl out of the water to begin the game anew.

I mention the spawning of the toads because it is one of the phenomena of spring which most deeply appeal to me, and because the toad, unlike the skylark and the primrose, has never had much of a boost from the poets. But I am aware that many people do not like reptiles or amphibians, and I am not suggesting that in order to enjoy the spring you have to take an interest in toads. There are also the crocus, the missel thrush, the cuckoo, the blackthorn, etc. The point is that the pleasures of spring are available to everybody, and cost nothing. Even in the most sordid street the coming of spring will register itself by some sign or other, if it is only a brighter blue between the chimney pots or the vivid green of an elder sprouting on a blitzed site. Indeed it is remarkable how Nature goes on existing unofficially, as it were, in the very heart of London. I have seen a kestrel flying over the Deptford gasworks, and I have heard a first-rate performance by a blackbird in the Euston Road. There must be some hundreds of thousands, if not millions, of birds living inside the four-mile radius, and it is rather a pleasing thought that none of them pays a halfpenny of rent.

As for spring, not even the narrow and gloomy streets round the Bank of England are quite able to exclude it. It comes seeping in everywhere, like one of those new poison gases which pass through all filters. The spring is commonly referred to as "a miracle," and during the past five or six years this worn-out figure of speech has taken on a new lease of life. After the sort of winters we have had to endure recently, the spring does seem miraculous, because it has become gradually harder and harder to believe that it is actually going to happen. Every February since 1940 I have found myself thinking that this time winter is going to be permanent. But Persephone, like the toads, always rises from the dead at about the same moment. Suddenly, toward the end of March, the miracle happens and the decaying slum in which I live is transfigured. Down in the square the sooty privets have turned bright green, the leaves are thickening on the chestnut trees, the daffodils are out, the wallflowers are budding, the policeman's tunic looks positively a pleasant shade of blue, the fish-monger greets his customers with a smile, and even the sparrows are quite a different color, having felt the balminess of the air and nerved themselves to take a bath, their first since last September.

Is it wicked to take a pleasure in spring, and other seasonal changes? To put it more precisely, is it politically reprehensible, while we are all groaning, under the shackles of the capitalist system, to point out that life is frequently more worth living because of a blackbird's song, a yellow elm tree in October, or some other natural phenomenon which does not cost money and does not have what the editors of the left-wing newspapers call a class angle? There is no doubt that many people think so. I know by experience that a favorable reference to "Nature" in one of my articles is liable to bring me abusive letters, and though the keyword in these letters is usually "sentimental," two ideas seem to be mixed up in them. One is that any pleasure in the actual process of life encourages a sort of political quietism. People, so the thought runs, ought to be discontented, and it is our job to multiply our wants and not simply to increase our enjoyment of the things we have already. The other idea is that this is the age of machines and that to dislike the machine, or even to want to limit its domination, is backward-looking, reactionary, and slightly ridiculous. This is often backed up by the statement that a love of Nature is a foible of urbanized people who have no notion what Nature is really like. Those who really have to deal with the soil, so it is argued, do not love the soil, and do not take the faintest interest in birds or flowers, except from a strictly utilitarian point of view. To love the country one must live in the town, merely taking an occasional weekend ramble at the warmer times of year.

This last idea is demonstrably false. Medieval literature, for instance, including the popular ballads, is full of an almost Georgian enthusiasm for Nature, and the art of agricultural peoples such as the Chinese and Japanese centers always round trees, birds, flowers, rivers, mountains. The other idea seems to me to be wrong in a subtler way. Certainly we ought to be discontented, we ought not simply to find out ways of making the best of a bad job, and yet if we kill all pleasure in the actual process of life, what sort of future are we preparing for ourselves? If a man cannot enjoy the return of spring, why should he be happy in a labor-saving Utopia? What will he do with the leisure that the machine will give him? I have always suspected that if our economic and political problems are ever really solved, life will become simpler instead of more complex, and that the sort of pleasure one gets from finding the first primrose will loom larger than the sort of pleasure one gets from eating an ice to the tune of a Wurlitzer. I think that by retaining one's childhood love of such things as trees, fishes, butterflies, and—to return to my first instance—toads, one makes a peaceful and decent future a little more probable, and that by preaching the doctrine that nothing is to be admired except steel and concrete, one merely makes it a little surer that human beings will have no outlet for their surplus energy except in hatred and leader-worship.

At any rate, spring is here, even in London, N. 1, and they can't stop you enjoying it. This is a satisfying reflection. How many a time have I stood watching the toads mating, or a pair of hares having a boxing match in the young corn, and thought of all the important persons who would stop me enjoying this if they could. But luckily they can't. So long as you are not actually ill, hungry, frightened, or immured in a prison or a holiday camp, spring is still spring. The atom bombs are piling up in the factories, the police are prowling through the cities, the lies are streaming from the loudspeakers, but the earth is still going round the sun, and neither the dictators nor the bureaucrats, deeply as they disapprove of the process, are able to prevent it.

Joseph Wood Krutch
From *The Twelve Seasons* (1949)
"The Day of the Peepers"

Hyla crucifer is what the biologists call him, but to most of us he is simply the Spring Peeper. The popularizers of natural history have by no means neglected him but even without their aid he has made himself known to many whose only wild flower is the daisy and whose only bird is the robin. Everyone who has ever visited the country in the spring has heard him trilling from the marsh at twilight, and though few have ever caught sight of him most know that he is a little, inch-long frog who has just awaked from his winter sleep. In southern Connecticut he usually begins to pipe on some day between the middle of March and the middle of April, and I, like most country dwellers, listen for the first of his shrill, cold notes.

Throughout the winter, neighbors who met in the village street have been greeting one another with the conventional question: "Is it cold enough for you?" Or, perhaps, if they are of the type which watches a bit more carefully than most the phenomenon of the seasons, they have been comparing thermometers in the hope that someone will admit to a minimum at least one degree higher than what was recorded "over my way." Now, however, one announces triumphantly: "Heard the peepers last night," and the other goes home to tell his wife. Few are High Church enough to risk a "Christ is risen" on Easter morning, but the peepers are mentioned without undue self-consciousness.

Even this, however, is not enough for me and I have often wondered that a world which pretends to mark so many days and to celebrate so many occasions should accept quite so casually the day when *Hyla crucifer* announces that winter is over. One swallow does not make a spring, and the robin arrives with all the philistine unconcern of a worldling back from his Winter at Aiken or Palm Beach. But the peeper seems to realize, rather better than we, the significance of his resurrection, and I wonder if there is any other phenomenon in the heavens above or in the earth beneath which so simply and so definitely announces that life is resurgent again.

We who have kept artificially warm and active through the winter act as though we were really independent of the seasons, but we forget how brief our immunity is and are less anxious than we might be if habit had not dulled our awareness. One summer which failed to arrive and we should realize well enough before we perished of hunger that we are only a little less at the mercy of the seasons than the weed that dies in October. One winter which lasted not six months but twelve and we should recognize our affinity with the insects who give up the ghost after laying the eggs that would never hatch if they did not lie chill and dead through the cold of a winter as necessary to them as warmth was to the males who fertilized and the females who laid them. We waited through the long period during which our accumulated supplies of food grew smaller and we waited calmly in a blind assurance that warmth would return and that nature would reawaken. Now, the voice of the peeper from the marsh announces the tremendous fact that our faith has been justified. A sigh of relief should go up and men should look at one another with a wild surprise. "It" has happened again, though there was nothing during the long months that passed to support our conviction that it could and would.

We had, to be sure, the waiting pages of our calendars marked "June," "July," and even, of all things, "August." The sun, so the astronomers had assured us, had turned northward on a certain date and theoretically had been growing stronger day by day. But there was, often enough, little in the mercury of our thermometers or the feel of our fingers to confirm the fact. Many a March day had felt colder than the milder days of February. And merely astronomical seasons have, after all, very little relation to any actual human experience either as visible phenomena or as events bringing with them concomitant earthly effects.

Not one man out of a hundred thousand would be aware of the solstices or the equinoxes if he did not see their dates set down in the almanac or did not read about them in the newspaper. They cannot be determined without accurate instruments and they correspond to no phenomena he is aware of. But the year as we live it does have its procession of recurring events, and it is a curious commentary on the extent to which we live by mere symbols that ten men know that the spring equinox occurs near the twenty-first of March to one who could give you even the approximate date when the peepers begin in his community; and that remains true even if he happens to be a countryman and even if he usually remarks, year after year, when they do begin.

It is true that the Day of the Peepers is a movable feast. But so is Easter, which—as a matter of fact—can come earlier or later by just about the same number of days that, on the calendar I have kept, separates the earliest from the latest date upon which *Hyla crucifer* begins to call. Moreover, the earliness or the lateness of the peepers means something, as the earliness or the lateness of Easter does not.

Whatever the stars may say or whatever the sun's altitude may be, spring has not begun until the ice has melted and life begun to stir again. Your peeper makes a calculation which would baffle a meteorologist. He takes into consideration the maximum to which the temperature has risen, the minimum to which it has fallen during the night, the relative length of the warmer and the colder periods, besides, no doubt, other factors hard to get down in tables or charts. But at last he knows that the moment has come. It has been just warm enough just long enough, and without too much cold in between. He inflates the little bubble in his throat and sends out the clear note audible for half a mile. On that day something older than any Christian God has risen. The earth is alive again.

The human tendency to prefer abstractions to phenomena is, I know, a very ancient one. Some anthropologists, noting that abstract design seems usually to come before the pictorial representation of anything in primitive man's environment, have said that the first picture drawn by any beginning culture is a picture of God. Certainly in the European world astronomy was the first of the sciences, and it is curious to remember that men knew a great deal about the intricate dance of the heavenly bodies before they had so much as noticed the phenomena of life about them. The constellations were named before any except the most obvious animals or plants and were studied before a science of botany or physiology had begun. The Greeks, who thought that bees were generated in the carcasses of dead animals and that swallows hibernated under the water, could predict eclipses, and the very Druids were concerned to mark the day on which the sun turned northward again. But the earliest of the sciences is also the most remote and the most abstract. The objects with which it deals are not living things and its crucial events do not correspond directly or immediately to any phenomena which are crucial in the procession of events as they affect animal or vegetable life.

Easter is an anniversary, and the conception of an anniversary is not only abstract but so difficult to define that the attempt to fix Easter used up an appalling proportion of the mental energy of learned men for many hundreds of years—ultimately to result in nothing except a cumbersome complexity that is absolutely meaningless in the end. Why should we celebrate the first Sunday after the first full moon on or after the twenty-first of March? What possible meaning can the result of such a calculation have? Yet even that meaningless definition of Easter is not really accurate. For the purpose of determining the festival, the date of the full moon is assumed to be, not that of the actual full moon, but that on which the full moon would have fallen if the table worked out by Pope Gregory's learned men had been—as it is not—really accurate. Even the relatively few men who remember the commonly given formula will occasionally find that they have missed their attempt to determine when Easter will be because they consulted a lay calendar to find the full moon instead of concerning themselves with the Epact and considering the theoretical ecclesiastical full moon rather than the actual one. How much easier it is to celebrate the Day of the Peepers instead, and how much more meaningful too! On that day something miraculous and full of promise has actually happened, and that something announces itself in no uncertain terms.

Over any astronomically determined festival, the Day of the Peepers has, moreover, another advantage even greater than the simplicity with which it defines itself or the actuality of its relation to the season it announces, for *Hyla crucifer* is a sentient creature who shares with us the drama and the exultation; who, indeed, sings our hosannahs for us. The music of the spheres is a myth; to say that the heavens rejoice is a pathetic fallacy; but there is no missing the rejoicings from the marsh and no denying that they are something shared. Under the stars we feel alone but by the pond side we have company.

To most, to be sure, Hyla is a *vox et praterea nihil*. Out of a thousand who have heard him, hardly one has ever seen him at the time of his singing or recognized him if perchance he has happened by pure accident to see squatting on the branch of some shrub the tiny inch-long creature, gray or green according to his mood, and with a dark cross over his back. But it was this tiny creature who, some months before, had congregated with his fellows in the cold winter to sing and make love. No one could possibly humanize him as one humanizes a pet and so come to feel that he belongs to us rather than—what is infinitely more important—that we both, equally, belong to something more inclusive than ourselves.

Like all the reptiles and the amphibians he has an aspect which is inscrutable and antediluvian. His thoughts must be inconceivably different from ours and his joy hardly less so. But the fact is comforting rather than the reverse, for if we are nevertheless somehow united with him in that vast category of living things which is so sharply cut off from everything that does not live at all, then we realize how broad the base of the category is, how much besides ourselves is, as it were, on our side. Over against the atoms and the stars are set both men and frogs. Life is not something entrenched in man alone, in a creature who has not been here so very long and may not continue to be here so very much longer. We are not its sole guardians, not alone in enjoying or enduring it. It is not something that will fail if we should.

Strangely enough, however, man's development takes him farther and farther away from association with his fellows, seems to condemn him more and more to live with what is dead rather than with what is alive. It is not merely that he dwells in cities and associates with machines rather than with plants and with animals. That, indeed, is but a small and a relatively unimportant part of his growing isolation. Far more important is the fact that more and more he thinks in terms of abstractions, generalizations, and laws; less and less participates in the experience of living in a world of sights, and sounds, and natural urges.

Electricity, the most powerful of his servants, flows silently and invisibly. It isn't really there except in its effects. We plan our greatest works on paper and in adding machines. Push the button, turn the switch! Things happen. But they are things we know about only in terms of symbols and formulae. Do we inevitably, in the process, come ourselves to be more and more like the inanimate forces with which we deal, less and less like the animals among whom we arose? Yet it is of protoplasm that we are made. We cannot possibly become like atoms or like suns. Do we dare to forget as completely as we threaten to forget that we belong rejoicing by the marsh more anciently and more fundamentally than we belong by the machine or over the drawing board?

No doubt astronomy especially fascinated the first men who began to think because the world in which they lived was predominantly so immediate and so confused a thing, was composed so largely of phenomena which they could see and hear but could not understand or predict and to which they so easily fell victim. The night sky spread out above them defined itself clearly and exhibited a relatively simple pattern of surely recurring events. They could perceive an order and impose a scheme, thus satisfying an intellectual need to which the natural phenomena close about them refused to cater.

But the situation of modern man is exactly the reverse. He "understands" more and more as he sees and hears less and less. By the time he has reached high-school age he has been introduced to the paradox that the chair on which he sits is not the hard object it seems to be but a collection of dancing molecules. He learns to deal, not with objects but with statistics, and before long he is introduced to the idea that God is a mathematician, not the creator of things seen, and heard, and felt. As he is taught to trust less and less the evidence of the five senses with which he was born, he lives less and less in the world which they seem to reveal, more and more with the concepts of physics and biology. Even his body is no longer most importantly the organs and muscles of which he is aware but the hormones of which he is told.

The very works of art that he looks at when he seeks delight through the senses are no longer representations of what the eye has seen but constructions and designs—or, in other words, another order of abstractions. It is no wonder that for such a one spring should come, not when the peepers begin, but when the sun crosses the equator or rather—since that is only a human interpretation of the phenomenon—when the inclined axis of the earth is for an instant pointed neither toward nor away from the sun but out into space in such a way that it permits the sun's rays to fall upon all parts of the earth's surface for an equal length of time. For him astronomy does not, as it did for primitive man, represent the one successful attempt to intellectualize and render abstract a series of natural phenomena. It is, instead, merely one more of the many systems by which understanding is substituted for experience.

Surely one day a year might be set aside on which to celebrate our ancient

Figure 11
Javan flying frog *(Rhacophorus reinwardti)*.
(From *The Riverside Natural History,* 1888, p. 341.)

loyalties and to remember our ancient origins. And I know of none more suitable for that purpose than the Day of the Peepers. "Spring is come!", I say when I hear them, and: "The most ancient of Christs has risen!" But I also add something which, for me at least, is even more important. "Don't forget," I whisper to the peepers; "we are all in this together."

Edwin Way Teale

From *Wandering Through Winter* (1965)
Chapter 26: "Audubon's Salamanders"

Down from the Appalachian Highland, down among the bare apple orchards of the Shenandoah Valley of Virginia, then out onto the Coastal Plain and across the swollen Potomac River, we made a leisurely advance the next day. We came to Washington in the early dusk. Swept like a chip in the torrent of traffic at the capital's perimeter, blinded by streams and whirlpools of headlights, we finally escaped to the north as though released from another flood. Before we settled down in a country motel, at the end of this initial day of March, we looked up at the sky. Both Leo and the variable star that has meant so much in the life of Leslie Peltier were hidden behind clouds.

The following morning began in a way no other dawn of ours had ever started. It commenced with a breakfast of hot-cake sandwiches—griddle cakes with ham between and a fried egg on top. Thus amply fortified against the cold, we whirled around Baltimore on a new expressway. Speeding along the multilane concrete of this bypass, we envisioned a future when a whole succession of such highways, like the growth rings of a city, may extend in ever wider curves around expanding urban centers.

North of Baltimore, we returned to a region filled with memories. Here the trails of our four journeys converged. When we stopped for the night at Havre de Grace, at the head of Chesapeake Bay, we slept in Cabin No. 7 at the Chesapeake Motor Court. It was part of all our travels with the seasons. In this identical cabin we had spent our first night when we had ridden south to meet the spring. Here we had stayed when we returned from our journey into summer. Here we had spent the night when we came home from the Pacific after crossing the continent through autumn. And now, in these latter days of winter, we returned to this same cabin once more.

The morning we left Havre de Grace and crossed the Susquehanna River, our wheels rolled over roads in portions of four states—Maryland, Delaware, New Jersey and Pennsylvania. They carried us past Valley Forge, where Washington's Continental Army spent the miserable winter of 1777–1778, and—five miles west of Norristown, near the community of Audubon, in Pennsylvania—they brought us down a long lane to the house at Mill Grove.

More than a century and a half have passed since Mill Grove formed John James Audubon's first home in America. Although he spent hardly two years here—about the same length of time that Thoreau lived at Walden—his name is linked to this rolling land along Perkiomen Creek almost as indissolubly as Thoreau's is to the pond he made famous. The art of each provided a kind of unique and permanent possession. During the comparatively short period Audubon stayed at Mill Grove, the direction of his life became fixed. There he began his study and painting of American birds. There he conducted the first bird-banding experiment made in America. There he met the girl he married; Lucy Bakewell, the most important influence in his life.

Today the ivy-covered brown fieldstone house, now more than two centuries old, is a museum holding relics of the adventurous life of the American Woodsman. In 1951, exactly 100 years after Audubon's death, the 120 acres of Mill Grove were added to the Montgomery County Park System as a sanctuary and permanent historical shrine. The curator, J. d'Arcy Northwood, a friend I had last seen at an osprey's nest on Cape May, welcomed us and installed us in his cottage down a side lane from the main house. Quickly we all changed into warm tramping clothes and started out on trails that wind for more than four miles over the hills and through the woods and along Perkiomen Creek.

Audubon used to refer to Mill Grove as "that blessed spot." Even on this chill, overcast winter day, its charm was evident. We walked among great trees, beech and hemlock and shagbark hickory. In only six weeks or so these wooded hillsides would be carpeted with the wildflowers of spring, with trillium and hepatica and Dutchman's-breeches and jack-in-the-pulpit. Then all the woodland paths would echo with the song of Audubon's favorite bird, the wood thrush.

The birds of Mill Grove probably are more varied today than they were when Audubon observed them. The land is more open, more diversified in habitat. Northwood's list for the area includes 155 species. Evening grosbeaks come in winter to feed on the seeds of the box elders. In summer, swifts nest inside the stone-and-brick chimney of the abandoned copper smelter far back in the woods. The cave on the Perkiomen where Audubon banded phoebe nestlings disappeared more than a century ago, but these birds, perhaps even descendants of the occupants of that historic nest, still return. For sixty years at least, phoebes have nested on a ledge under the back porch at Mill Grove.

Although the Perkiomen is called a creek, it is larger than the River Thames during most of its flow through England. We watched mallard and widgeon and mergansers swimming on the open water below the dam where the old mill once stood. It was on the Perkiomen that Audubon saw a bald eagle pursuing fish on foot in shallow water. And it was here, one day in winter, that he nearly lost his life when the ice broke under him. A few days before we arrived, Northwood had observed eleven common mergansers fishing together, swimming upstream almost abreast, the line stretching from bank to bank.

By bracing ourselves against first one tree and then another, we worked down a zigzag trail along the face of a fifty-foot cliff of reddish rock. It drops almost perpendicularly to the bank of Mine Run near the spot where this smaller stream empties into the Perkiomen. At its base the wall is pierced by the openings of two ancient mine tunnels. We peered into the dark interiors. The timbers have rotted away. Cave-ins have occurred. From each opening flows the clear water of springs that miners long ago encountered deep within the rock of the cliffs. Even on the bitterest days of winter, the temperature of this flow remains at about fifty-two degrees F. Such spring water and the damp darkness of the old tunnels comprise the winter world of slender, moist-skinned, cold-blooded creatures we especially hoped to see. For Mill Grove provides a sanctuary for long-tailed salamanders as well as for wildflowers and birds.

When autumn nights grow chill, these winter-born amphibians migrate under the cover of darkness from the vicinity of Mine Run and Perkiomen Creek to the underground streams that flow through the two man-made caverns of the cliff. There, mainly in November and December, they lay their eggs. It is usually in January, at a time when the outer world is frozen and blanketed with snow, when life seems at its lowest ebb, that the eggs hatch and existence begins for the larvae of each new generation of long-tailed salamanders. For this species, the period of reproduction is over at a time when, for most other salamanders, the breeding season is just beginning.

In 1952 Charles E. Mohr, in his continuing studies of cave life, made extensive observations of the salamanders of the Mill Grove tunnels. At frequent intervals he kept track of the activity and movement of the amphibians. Nature, he found, had marked individuals in a kind of code using lines and spots. No two of the amphibians exhibited precisely the same pattern of markings. By recording the arrangement of black splotches on the yellow to orange-red ground color of the skin, he was able to recognize individuals when he encountered them later on. At times, he discovered, the salamanders would remain in the same place day after day. At other times, they would wander along the floor of the tunnel as far as 300 feet from the entrance. The studies which Mohr made at Mill Grove provided the first precise record of the movement of long-tailed salamanders in their winter home.

As its name implies, this species is distinguished by an abnormally long tail, often half again the length of the body. Its scientific name, as well as its tail, is long—*Eurycea longicauda longicauda*. At Mill Grove it is approaching the northern limit of its range, which extends just above the Pennsylvania line into New York. Adults usually measure from four to six and a quarter inches in length. The record for this species is seven and one-eighth inches. Nowhere else in the world are salamanders as varied and numerous as in the eastern United States. The Appalachian plateau is the world center of abundance for these shy and secretive creatures.

In only two places at Mill Grove, about a mile apart, are long-tailed salamanders known to spend the winter and reproduce. One is, as we have seen, the twin tunnels beside Mine Run. The other is in an artesian well whose clear, cold water flows from Triassic shale in the basement of Northwood's cottage. At the tunnels the creatures enter directly from the out-of-doors, but to reach the artesian well the procedure is not so simple. First the salamanders follow upstream a brooklet that descends for a quarter of a mile to Perkiomen Creek. Then they enter the lower end of a small iron pipe that, day and night the year around, carries away the overflow from the artesian spring. Fighting against the rush of the water, they work their way up this pipe for 100 feet. At its upper end they emerge into the covered concrete box sunk into the basement floor, within which the water of the spring wells up continually.

Back at the cottage we shed our sweaters and woolen gloves and earflap caps and descended to the well. But here, as at the tunnels, we saw no salamanders. We had arrived just too late. By early March the larvae and the adults have already begun spreading away from their winter quarters. To see them, I returned a few weeks earlier the following year. Then, as I leaned over the well, the beam of Northwood's flashlight ran along the concrete at the water's edge. One after another, it glinted on the moist, shining skins of half a dozen salamanders. Most of them clung to the rough interior of the well box just above the water line. One was almost hidden where it had squeezed itself into a crack in the concrete. Others were resting on stones or riding on the floating pieces of wood that, each autumn, Northwood places in the well. All were very thin. It had been months since they had eaten. Moreover, the females now had deposited all their eggs.

More than 100 of these gelatinous spheres, each about the size of a pea, were still unhatched in the well. Most were clinging to flat stones or to the undersides of floating bits of wood. At one place where a spray of fine, feathery box-elder rootlets, after reaching the well through a small crack in the concrete, swayed back and forth in the currents of the spring, a score or more of the globes were thickly strewn throughout the mass. I had arrived, I found, just in the nick of time. For, even as I bent closer, the most dramatic event of the salamander's winter was taking place. The eggs, one by one, were disappearing as the larvae hatched and swam away.

I fished one almost transparent egg from the spring and held it in the water cupped within my hand. Inside I could see the bowed form of the gilled, fishlike larva. It resembled a quarter moon. As I watched, the larva wriggled, the shell of the globe seemed to dissolve, and a new salamander was launched into its watery world.

Within the well, as soon as a larva hatched and swam away, it vanished. It became invisible against the darker bottom of the spring. On emerging from the egg, the hatching salamander is about three-quarters of an inch long. Before it disappears down the overflow pipe by early March, it has nearly doubled its size and has developed a yellowish tail.

The water in which salamander life begins in this artesian well remains, the year around, at almost a constant 50 degrees F. Like all cold-blooded creatures, salamanders are made less lively by cold and more active by warmth. When Northwood caught one adult in his hand and we were bending close to examine it, it warmed up in contact with his skin. In a sudden leap it sailed downward for a yard or more and splashed back into the water of the well.

About these soft-bodied, moist-skinned creatures there is a kind of helpless innocence and charm. They seem to have crept directly from the pages of Charles Kingsley's childhood classic, *The Water Babies.* And young salamanders are, in truth, water babies. With their tiny tufts of gills, they can live only under water. The air-breathing adults—usually found secreted in such places as under rotting logs or beneath streamside stones—also require moist conditions to survive. On several occasions Northwood has come upon the dry bodies of wanderers that have strayed from the well onto the concrete of the basement floor and, unable to return, have become desiccated and died.

In the main, salamanders avoid direct sunlight and tend to prefer cool to warm weather. They are often abroad at times of heavy mist and after thunderstorms. Some species are able to remain active even in icy water close to the freezing point. Although at first glance salamanders seem frail, defenseless, vulnerable to their foes, these creatures have endured for more than 100,000,000 years. Moreover, for them the loss of a leg or tail is no great matter. Unlike the higher animals, they grow another in its place.

Each winter the colony returns to the flowing well in the basement of the cottage. How did it become established? How did its annual migration up the brooklet and the long iron pipe commence? As Nellie and I sat with Northwood on that evening of our winter trip, we discussed these questions. His logical explanation was that the spring long antedated the cottage. The salamanders had been used to following the brooklet to its source and there laying their winter eggs. When the house was built and the iron pipe installed, they continued moving up against the flow until they reached the well.

While we talked, the barometer was falling. Later, when we returned from examining the elephant folio and the many other items of interest at the house where Audubon once lived, we walked down the lane in a rising wind. The thermometer was going down. Gale winds and heavy snow were predicted on the radio before we went to bed. When we awoke in the crepuscular light of a leaden dawn, we could see branches lashing in the wind and almost horizontal lines of snow being hurled past our bedroom window. A howling blizzard raged outside. The barometer continued to fall.

As we ate poached eggs and buttered toast for breakfast, we watched chickadees and tufted titmice being tossed about by the gusts at a feeding tray beyond the window. We debated what to do. The sensible thing was to stay put until the storm was over. But, as Audubon long ago observed: "Time is ever precious to the student of nature." The days of our winter trip were running out. The lane leading from the cottage was narrow and steep; the road it joined would be one of the last to be plowed out. We might be snowed in for days. If we could get out of the Pennsylvania hills onto the flat land of New Jersey, we could be sure of being able to go on as soon as the storm had ended. We looked out at the snow. Although it was building up fast, it was no more than a couple of inches deep. We should have a few hours. We decided in favor of the calculated risk. With considerable misgivings all around, we said goodbye to our friend. Then we headed our white car out into the white gale. When we looked back from the road, Audubon's first American home was already lost in the scudding snow.

Annie Dillard
From *Pilgrim at Tinker Creek* (1974)
Chapter 1: "Heaven and Earth in Jest"

A couple of summers ago I was walking along the edge of the island to see what I could see in the water, and mainly to scare frogs. Frogs have an inelegant way of taking off from invisible positions on the bank just ahead of your feet, in dire panic, emitting a froggy "Yike!" and splashing into the water. Incredibly, this amused me, and, incredibly, it amuses me still. As I walked along the grassy edge of the island, I got better and better at seeing frogs both in and out of the water. I learned to recognize, slowing down, the difference in texture of the light reflected from mudbank, water, grass, or frog. Frogs were flying all around me. At the end of the island I noticed a small green frog. He was exactly half in and half out of the water, looking like a schematic diagram of an amphibian, and he didn't jump.

He didn't jump; I crept closer. At last I knelt on the island's winterkilled grass, lost, dumbstruck, staring at the frog in the creek just four feet away. He was a very small frog with wide, dull eyes. And just as I looked at him, he slowly crumpled and began to sag. The spirit vanished from his eyes as if snuffed. His skin emptied and drooped; his very skull seemed to collapse and settle like a kicked tent. He was shrinking before my eyes like a deflating football. I watched the taut, glistening skin on his shoulders ruck, and rumple, and fall. Soon, part of his skin, formless as a pricked balloon, lay in floating folds like bright scum on top of the water: it was a monstrous and terrifying thing. I gaped bewildered, appalled. An oval shadow hung in the water behind the drained frog; then the shadow glided away. The frog skin bag started to sink.

I had read about the giant water bug, but never seen one. "Giant water bug" is really the name of the creature, which is an enormous, heavy-bodied brown beetle. It eats insects, tadpoles, fish, and frogs. Its grasping forelegs are mighty and hooked inward. It seizes a victim with these legs, hugs it tight, and paralyzes it with enzymes injected during a vicious bite. That one bite is the only bite it ever takes. Through the puncture shoot the poisons that dissolve the victim's muscles and bones and organs—all but the skin—and through it the giant water bug sucks out the victim's body, reduced to a juice. This event is quite common in warm fresh water. The frog I saw was being sucked by a giant water bug. I had been kneeling on the island grass; when the unrecognizable flap of frog skin settled on the creek bottom, swaying, I stood up and brushed the knees of my pants. I couldn't catch my breath.

Of course, many carnivorous animals devour their prey alive. The usual method seems to be to subdue the victim by downing or grasping it so it can't flee, then eating it whole or in a series of bloody bites. Frogs eat everything whole, stuffing prey into their mouths with their thumbs. People have seen frogs with their wide jaws so full of live dragonflies they couldn't close them. Ants don't even have to catch their prey: in the spring they swarm over newly hatched, featherless birds in the nest and eat them tiny bite by bite.

That it's rough out there and chancy is no surprise. Every live thing is a survivor on a kind of extended emergency bivouac. But at the same time we are also created. In the Koran, Allah asks, "The heaven and the earth and all in between, thinkest thou I made them *in jest*?" It's a good question. What do we think of the created universe, spanning an unthinkable void with an unthinkable profusion of forms? Or what do we think of nothingness, those sickening reaches of time in either direction? If the giant water bug was not made in jest, was it then made in earnest? Pascal uses a nice term to describe the notion of the creator's, once having called forth the universe, turning his back to it: *Deus Absconditus.* Is this what we think happened? Was the sense of it there, and God absconded with it, ate it, like a wolf who disappears round the edge of the house with the Thanksgiving turkey? "God is subtle," Einstein said, "but not malicious." Again, Einstein said that "nature conceals her mystery by means of her essential grandeur, not by her cunning." It could be that God has not absconded but spread, as our vision and understanding of the universe have spread, to a fabric of spirit and sense so grand and subtle, so powerful in a new way, that we can only feel blindly of its hem. In making the thick darkness a swaddling band for the sea, God "set bars and doors" and said, "Hitherto shalt thou come, but no further." But have we come even that far? Have we rowed out to the thick darkness, or are we all playing pinochle in the bottom of the boat?

Chapter 7: "Spring"

In April I walked to the Adams' woods. The grass had greened one morning when I blinked; I missed it again. As I left the house I checked the praying mantis egg case. I had given all but one of the cases to friends for their gardens; now I saw that small black ants had discovered the one that was left, the one tied to the mock-orange hedge by my study window. One side of the case was chewed away, either by the ants or by something else, revealing a rigid froth slit by narrow cells. Over this protective layer the ants scrambled in a frenzy, unable to eat; the actual mantis eggs lay secure and unseen, waiting, deeper in.

The morning woods were utterly new. A strong yellow light pooled between the trees; my shadow appeared and vanished on the path, since a third of the trees I walked under were still bare, a third spread a luminous haze wherever they grew, and another third blocked the sun with new, whole leaves. The snakes were out—I saw a bright, smashed one on the path—and the butterflies were vaulting and furling about; the phlox was at its peak, and even the evergreens looked greener, newly created and washed.

Long racemes of white flowers hung from the locust trees. Last summer I heard a Cherokee legend about the locust tree and the moon. The moon goddess starts out with a big ball, the full moon, and she hurls it across the sky. She spends all day retrieving it; then she shaves a slice from it and hurls it again, retrieving, shaving, hurling, and so on. She uses up a moon a month, all year. Then, the way Park Service geologist Bill Wellman tells it, "long about spring of course she's knee-deep in moon-shavings," so she finds her favorite tree, the locust, and hangs the slender shavings from its boughs. And there they were, the locust flowers, pale and clustered in crescents.

The newts were back. In the small forest pond they swam bright and quivering, or hung alertly near the water's surface. I discovered that if I poked my finger into the water and wagged it slowly, a newt would investigate; then if I held my finger still, it would nibble at my skin, softly, the way my goldfish does—and, also like my goldfish, it would swim off as if in disgust at a bad job. This is salamander metropolis. If you want to find a species wholly new to science and have your name inscribed Latinly in some secular version of an eternal rollbook, then your best bet is to come to the southern Appalachians, climb some obscure and snakey mountain where, as the saying goes, "the hand of man has never set foot," and start turning over rocks. The mountains act as islands; evolution does the rest, and there are scores of different salamanders all around. The Peaks of Otter on the Blue Ridge Parkway produce their own unique species, black and spotted in dark gold; the rangers there keep a live one handy by sticking it in a Baggie and stowing it in the refrigerator, like a piece of cheese.

Newts are the most common of salamanders. Their skin is a lighted green, like water in a sunlit pond, and rows of very bright red dots line their backs. They have gills as larvae; as they grow they turn a luminescent red, lose their gills, and walk out of the water to spend a few years padding around in damp places on the forest floor. Their feet look like fingered baby hands, and they

From *Pilgrim at Tinker Creek* (1974)   145

walk in the same leg patterns as all four-footed creatures—dogs, mules, and, for that matter, lesser pandas. When they mature fully, they turn green again and stream to the water in droves. A newt can scent its way home from as far as eight miles away. They are altogether excellent creatures, if somewhat moist, but no one pays the least attention to them, except children.

Once I was camped "alone" at Douthat State Park in the Allegheny Mountains near here, and spent the greater part of one afternoon watching children and newts. There were many times more red-spotted newts at the edge of the lake than there were children; the supply exceeded even that very heavy demand. One child was collecting them in a Thermos mug to take home to Lancaster, Pennsylvania, to feed an ailing cayman. Other children ran to their mothers with squirming fistfuls. One boy was mistreating the newts spectacularly: he squeezed them by their tails and threw them at a shoreline stone, one by one. I tried to reason with him, but nothing worked. Finally he asked me, "Is this one a male?" and in a fit of inspiration I said, "No, it's a baby." He cried, "Oh, isn't he *cute!*" and cradled the newt carefully back into the water.

No one but me disturbed the newts here in the Adams' woods. They hung in the water as if suspended from strings. Their specific gravity put them just a jot below the water's surface, and they could apparently relax just as well with

Figure 12
Hellbender *(Cryptobranchus alleganiensis)*.
(From *The Riverside Natural History,* 1888, p. 310.)

lowered heads as lowered tails; their tiny limbs hung limp in the water. One newt was sunning on a stick in such an extravagant posture I thought she was dead. She was half out of water, her front legs grasping the stick, her nose tilted back to the zenith and then some. The concave arch of her spine stretched her neck past believing; the thin ventral skin was a bright taut yellow. I should not have nudged her—it made her relax the angle of repose—but I had to see if she was dead. Medieval Europeans believed that salamanders were so cold they could put out fires and not be burned themselves; ancient Romans thought that the poison of salamanders was so cold that if anyone ate the fruit of a tree that a salamander had merely touched, that person would die of a terrible coldness. But I survived these mild encounters—my being nibbled and my poking the salamander's neck—and stood up.

Robert Michael Pyle

From *Wintergreen: Listening to the Land's Heart* (1986)
Prologue

At any time of the year and in any weather, my bedroom window frames a green and pleasant country scene. Halfway open, it makes a Kodachrome slide of the bucolic valley below, bordered by white sashes and molding. Timbered hills tumble down to a floodplain pasture valley, bounded below by a limpid river, itself spanned by an old gray covered bridge. Holstein cattle spot the meadows with black and white, and for half the year swallows pock and streak the broad skies.

I came to rural Wahkiakum County, Washington, at the end of the 1970s. This old homestead, known as Swede Park, represented release from the stress and distraction of the city. Gray's River, the stream and the town, seemed to provide the peaceful setting I needed in order to write full-time. The surrounding Willapa Hills, little studied by biologists, offered a fruitful field of exploration for a too-long-urban naturalist.

For the first few years, I maintained a pattern of commuting to Cambridge, England, for work that supported my writing habit when I was at home. By the time I settled in at Swede Park for good, I had lived in Great Britain four of the past ten years. So similar seemed the British Isles and the Pacific Northwest in some respects that I felt continually disoriented at first. To be sure, the differences are so obvious in terms of hardwoods versus conifers, maturity of landscapes and settlements, culture, architecture, and antiquities, that I soon realized the one will never be the other. Yet certain features remained evocative—chiefly, the climate and the colors. That it rains a lot in England and Washington comes as no surprise to anyone; nor that both mean green, green countryside. Rain makes green, as in Ireland and Oregon also. Both British and Northwest landscapes generate a solid and similar green gestalt. So in my travels back and forth I felt comfortable, jet lag and culture shock alike buffered by the soft mental bed of moss and grass that lay at either end.

William Blake described England as "a green and pleasant land." Many authors have since agreed. W. H. Hudson expanded on the idea when, in *Afoot in England,* he wrote of the River Otter in Devon as "the greenest, most luxuriant [place] in its vegetation . . . where a man might spend a month, a year, a lifetime, very agreeably, ceasing not to congratulate himself on the good fortune which first led him into such a garden." Such are my feelings about Gray's River. . . .

Chapter 8: "Waterproof Wildlife"

If the rain repels humans and butterflies, it well suits the newts and certain other animals. I call these the waterproof wildlife of Willapa. For those less well prepared, the point is adaptability rather than true suitability. If a duck takes to water like a duck, then a sapsucker may do so somewhat more reluctantly but do so nonetheless, rather than abandon a fine, sappy old orchard in a downpour. The key word is tolerance, and cheerfulness has very little to do with it. Whether naturally suited to water or not, an organism must in some way come to terms with the fact of much wetness if it is to survive in Willapa.

Actually, it's easy enough for us. As David Brower said in the narration to a Sierra Club film on the North Cascades, we do well to remember that skin, after all, is waterproof. With a bit of Gore-Tex or oilskin, and a few adjustments to prevent hypothermia and severe depression, humans adapt to more than one hundred inches of rainfall annually quite readily.

It can be more difficult for other animals. You might think, for instance, that slugs in their shiny damp viscidity were better at being wet than we are. But slug skin is *not* waterproof. These walking water balloons can drown or desiccate quite easily if conditions exceed their tolerance for moisture or drought. So, while slugs prosper in the rain world, they do so through sensitive physiological and behavioral adaptations to moisture excess. They are able to find the damp and stand it with equal proficiency.

Quite a few creatures have managed the task and some have mastered it to the point where we really can call them waterproof wildlife. Among the best are the newts and their relatives, the Amphibia. Like the fish from which they evolved, amphibians retain their physiological dependency upon water. In the larval stage most possess gills, either internal or external, and therefore are necessarily aquatic. Adults may be either with or without lungs, in the latter case breathing through their skin. That trick requires thin, moist skin, able to lose or gain water readily, and, consequently, damp conditions. "Drinking" too is by absorption through the skin. Clearly, these creatures find the water fine.

So amenable do amphibians find life in the wet world of Willapa that we can quite properly call it salamander land. According to Dr. Dennis Paulson, zoologist at Seattle's Burke Museum, Wahkiakum County is probably the center of diversity for Washington salamanders. This means that more species occur here than anywhere else in a state known for its salamander diversity. A glance at the field guides shows more species in the American Southeast; but the Southeast is essentially subtropical, and many groups have proliferated there in the ancient, moist hardwood forests and coastal plains. For a young, far-from-tropical region, our salamander fauna is impressive.

In little more than casual searching, my family and I have found red-backed, Olympic, Dunn's, northwestern, and long-toed salamanders in local woods and streams, as well as the ubiquitous rough-skinned newts. Ensatina, Pacific Giant, Van Dyke's, and the others will probably show up in good time. We search in the classic places—under logs and stones, alongside clear streams, on mossy forest floors, often with good results. Salamanders make good terrarium subjects, though we usually release them before long in our own stream, whose ravine contains all the above-mentioned microhabitats. So far, two or three species have "taken" or else shown up on their own.

With their aquatic breeding habits and larval stages, and the need to keep their skins moist, salamanders perfectly suit the land of much rainfall—or vice versa. Suitable habitats, as we have found, abound. Sometimes anthropogenic settings suffice for, or even attract, the amphibians. A swimming pool valve-box on Orcas Island proved a treasure chest of trapped newts and salamanders, which we released to a nearby pond. Recently a young northwestern salamander appeared in our stone cellar, having arrived through the walls like the winter spurtings of the floodwater, or up the sump-pump hole, but now in late summer nearly dried to death. It too was saved to enrich our pond, once for trout and now a sanctuary for salamanders.

That find put me in mind of last March, when the spring rains brought many amphibians in search of one another out of hiding and onto the slick roads. One can see the attraction—the flat surface of the highway would be suitably moist, yet clear of obstacles to the search for a mate. Now and again the courtship hunt would even be illuminated by the headlights of helpful drivers. Actually not so helpful: we found it easy to miss most of the 'manders with a little care—"We brake for newts" came to mind as a suitable sticker—if not the moving targets of frogs. Yet, sadly, the roadway lay littered with flatter-than-usual salamanders. So the adoption of the public right-of-way as a rendezvous proves maladaptive after all.

In particular, we noted the massive carcasses of northwestern salamanders. Two of these, both nipped clinically by the head and otherwise intact, we collected for examination at home. The postmortem proved fascinating. John McPhee, in his essay "Travels in Georgia," has written of the pleasures of roadkill cookery. I find that D.O.R. (dead-on-road) animals, if fresh and intact, furnish elegant opportunities for highly instructive dissection. Thus may the lessons of college morphology classes continue without having to kill the subjects, and thus may poor road-kills be appreciated, even by those disinclined to consume them.

They turned out to be a male and a female, the former measuring 21.5 cm and weighing 30 gms; the latter, 24 cm and 60 gm. That made the female over nine and one-half inches and two ounces, the size of many a respectable mammal, a truly imposing animal. Though just an inch longer than the male, the female was twice his weight. The difference seemed to be accounted for within by elaborate, convoluted, paired masses of white tubes, stretching to more than 50 cm. I took these for ovaries. The actual eggs, consisting of seventy-five or so in each of two jade-green clusters, lay in yellow sacs at the ends of the oviducts. What surprised me was that the other salamander, while possessing the swollen cloaca of the male and apparent testes, also had the ovaries—though much smaller and less well developed and with no sign of eggs.

Perhaps I should not have been surprised after all, since hermaphroditism is well known among amphibians. In *The Sex Life of the Animals,* Herbert Wendt described how the sex of many young frogs is indeterminate at first. Among obvious males and females, one finds numerous examples of AC/DC adolescents bearing juvenile gonads of both sexes. Apparently, these may eventually go either way. Most of the hermaphrodites will mature into males, some into females. Salamanders are not mentioned, but my sample of two road-kills—one definitely female, one ambiguous—seems to show that these creatures too may be sexually ambivalent in youth.

I suspect that if I knew all the secrets of salamander sex, inside and out, I should have a capital tale. How, for example, do they locate one another in the dark nights of their roamings? It isn't likely that the shine of their liquid eyes and glistening skin in the dangerous headlights helps at all. The scent of pheromones released in the waters of their destination probably attracts salamanders to one another, at least those that have made up their minds and bodies about which sex they intend to be. Males release their sperms in a gelatinous spermatophore at the proper time as determined by sometimes elaborate courtship procedures; the females then collect this gift with their cloacas. I'm sure we lack many facts, among them the nature of satisfactions gained, if any, through such a chaste form of fertilization. One assumes that, the act completed, hormones allow the mating frenzy to subside, whether the union involves copulo or not; and that must be a form of satisfaction. How little we really know of different strokes.

One thing we do know is something of the way in which these big amphibians repel predators (as they must, for they offer such easy prey in their precourtship perambulations on land). These rubbery animals possess thick, spongy glands behind their eyes, along their sides, and up and down their muskratlike flattened tails. When I squeezed these with the dissecting needle, they fairly oozed a toxic fluid much like milkweed juice. The thick, white stuff stood by in such ready quantity that, from one pressed gland, it squirted into my face, several inches away. I hurried to wash it off. My fingertips were roughened and tender from handling the copious, milky venom. Who would eat such a beast?

Our rough-skinned newts have their venom too, which they advertise with their bright red bellies—a form of self-promotion known as warning coloration. Many brightly colored, distasteful insects, like the monarch butterfly, do this. But for the population to acquire protection, the potential predators have to learn to avoid them through initial unpleasant contact. Hence, the famous photograph of a barfing blue jay in Lincoln Brower's 1969 *Scientific American* article on ecological chemistry. The naive jay, having tried one monarch and found it extremely unpleasant, has been educated: it represents no further threat to monarchs. Look-alike viceroys, though palatable, gain protection by mimicking monarchs. A news item in an Oregon paper a couple of years ago suggested that at least one logger had less sense than a blue jay. Dared to drink a newt in his beer, he did so—and died three hours later from acute toxic reaction. No more red-bellies in the beer in that tavern, I'll wager. The ploy worked for the local population of newts if not for the test case.

Another western amphibian known for its venomous glands, the northern toad (of the lovely name *Bufo boreas*), should by all reckoning proliferate throughout the moist, green land of Willapa. But, as we know from Thornton Burgess and Kenneth Grahame, old Mr. Toad is a willful animal. Both writers were as much sharp watchers as they were clever storytellers, and they knew an animal's nature before investing it with character. As Rat and Mole found, Toad just can't be trusted to do what you expect him to do. And so it is here. I can't find a toad for the life of me, nor discern any limiting factors for their living here. Surely not the Rot Factor that I believe discourages butterflies, for they thrive in the moldiest spots. Can't be cold, for I've met them high in the Colorado alpine. Yet they are just not about as they are elsewhere. I have never seen a toad at Swede Park, apparently as fine a Toad Hall as ever there was; and I know of only one or two places where toads have been spotted in the Willapa Hills by reliable observers. The fact that from time to time they get trapped in the ponds at the Naselle Salmon Hatchery proves that they do in fact occur in the area; but why so sparse or retiring? I wonder.

Frogs, on the other hand, abound. Of course, frogs tend to be even more moisture-loving than toads, so their predominance here should not be too surprising. Both Pacific tree frogs and red-legged frogs (more euphonious names: *Hyla regilla, Rana aurora*) appear in the damp recesses of the woods in great numbers. Both take to water, of course, to breed. The spring-green tree frogs begin calling in the rushy meadow-swamps in February or March and carry on their nocturnal disquisition for months. It is always the same, each evening of each spring: first a single, tentative "rigit," becoming hortatory as more voices arise, finally a full-blown shout as inflated trills join in concert; then tailing off in exhaustion, ennui, or simple satiety. But before the croakers rest, each a jade and jaded voice box with a thousand-watt amplifier, they've given background music to an entire season.

Red-legs issue their quiet croaks from ponds in the woods, often calling underwater if the books are to be believed. Later in the year they show up in the forest. Some, almost as large as modest bull-frogs, startle the prowler of the fern banks by stirring far from any water. How, in a dry summer such as this just passed, do they remain moist? By the dry time the tree frogs too have shifted position. When the water withdraws from the ponds and the rushy bottoms, they take to the trees and bushes. And by August or so, all we hear is the occasional katydidlike croak from an oak, or the feeble "braack" of a leaf-mold lurker. The ponds and river backwaters boil with polliwogs, and soon the tiniest frogs you ever saw—bright green mites with raccoon masks and golden eyes or mottled brown hoppers with strong pink thighs—populate the woods to wait for winter wet and yet another spring.

Bullfrogs, introduced, float in shallow lakes with their huge eyes poking out: joke frogs. Other native species we have yet to find in their favored waters. One of these species recorded for the hills has shown up only as two tadpoles. Ed Maxwell, watchful keeper of the Naselle Salmon Hatchery, who spotted most of the local toads I know of, also found a pair of immature tailed frogs, one each in the Naselle and Gray's rivers.

The tailed frog *(Ascaphus truei)* has to be regarded as one of our most interesting organisms. Its name originates in the caudal flap that covers the genitalia of the male. This "tail" is actually a copulatory organ that facilitates a kind of internal fertilization—a sexual practice engaged in by no other frogs. Such an adaptation probably arose to permit mating in the swift streams occupied by the species. Without it, the spermatophore would be swept away on the current more often than not.

The tailed frog's distribution is as interesting as its sexual structure. Restricted to the Pacific Northwest *sensu latu,* it occurs in many separate streamside populations in California, Oregon, Idaho, Montana, Washington, and British Columbia. Yet its only close relatives are three species of terrestrial frogs in New Zealand. Lacking the genital "tail," these frogs nonetheless share sufficient traits with *Ascaphus* to make the relationship clear.

So how does a frog-family founder get from New Zealand to Oregon, or vice versa, and why doesn't it show up elsewhere in between? Consulting my old college notes from the late Professor Frank Richardson's excellent class in zoogeography, I find that these long-lost cousins are prime examples of relicts. (Relicts are species left behind on outposts when their formerly more widespread kin become extinct in between.) I read, too, that the tailed frog and its relatives are considered to be the oldest, most primitive family of frogs. So presumably they evolved long ago, whether in the South Pacific, the Pacific Northwest, or somewhere entirely different; dispersed; then died out in all but the two regions.

Frogs get around, like other animals, either under their own steam or by rafting across seas in storm-tossed bundles of wrack. Continental drift may have affected their earlier whereabouts. They drop out in the intervals due to all the catastrophic or gradual changes that bring about extinctions everywhere. The resulting pattern illustrates disjuncture, whereby once-widespread species have withdrawn over a long period to isolated fragments of their former range. Disjunct relicts are not infrequent in the floras and faunas, but few examples are quite as dramatic as the tailed frog and its New Zealand relatives.

Having failed to find it here for years, I suspected that isolation might have excluded the tailed frog from the Willapa Hills; or else, the fact that it seems to favor cold-water streams, whereas most of ours are relatively warm due to low altitudes and lack of glaciers or snowmelt upstream. But Ed's discovery would seem to prove that the animal occurs in Willapa. Even so, he might have missed it but for another remarkable trait of the tailed frog: an anchor against the current.

The tadpole's mouth is as interesting as the opposite end of the adult. Possessed of a sucker, it clings to stones in the rapid streams where the species lives. When Ed worked at the Klickitat Hatchery in the southern Cascades, he noticed large numbers of tailed-frog larvae (his co-workers thought they were leeches) attached to the cement side of fish ponds, where they apparently fed on algae and microorganisms. That experience keyed Ed to watch for the unique polliwogs when he moved to other hatcheries. After he'd found the first one in the Gray's River, we figured it just might have been an introduction, since Klickitat fish had been moved to Gray's River, and the tadpoles could have hitched along. But Ed's duplicate discovery in the Naselle River clinches the case: tailed frogs do dwell in the Willapa Hills.

Still, in order to find the adults and finally see the tail of the male, we shall, I suppose, be obliged to spend still more time alongside the rivers of these hills, perhaps at night, watching for the eye-shine of the nocturnal frogs, or overturning stones by day. No penalty this, for we love to do such things. Finding the tailed frog, which I have never seen, would be a great satisfaction. And in such pursuits, the pleasure lies largely in the search itself.

I remember a day spent in Mount Rainier National Park with my friend Noble Proctor, a New England naturalist intimately associated with the Roger Tory Peterson Institute. Noble had never seen *Ascaphus* either, and his blood was up for finding it. Avidly, we began turning cobbles in the Nisqually River. Eventually we were forced to give up as our hands stiffened in the glacial meltwater and the day's light ducked out. But the search added one more shining coin to our joint treasury of Mount Rainier memories, and we know we shall try again.

Meanwhile, my wife, Thea, and I intend to join the Maxwells in seeking tailed frogs in the rivers of these hills. I can hear the tavern-talk of more conventional hunters, bringing back their elk or deer, declaiming upon the sight they saw: " . . . and here were these people hunting some goddamned, fancypants FROGS!" To each his own.

Ann Haymond Zwinger

From *The Mysterious Lands: A Naturalist Explores the Four Great Deserts of the Southwest* (1989)
Chapter 23: "Of Spadefoot Toads and Twilight"

I kick a bootful of snow into the air. On this January day in the Great Salt Desert, the snow falls like a fine, soundless, glittering shower of fireworks. There are no horizons, only thick fog, wrapped like an ermine cape across the shoulders of the land.

All the way down the hill, the snow glistens, paved with ice crystals that rise an inch long and half an inch wide, stacked above the surface like feathers.

They are hexagonal structures, striped with clear white lines. I run my hand across the surface and hear a subdued wind-chimes tinkle. Terry Williams says they sound like African thumb pianos. She calls them *Qulu,* an Eskimo term that defines the precise conditions necessary to produce such a magic display of ice crystals.

The temperature is 19 degrees F. The only world that exists is the world at my fingertips, fine and brittle, a vast silence of freezing, swirling mist. The tallest objects in view are the dried sunflower and ragwort stalks leaning awry like untended fence posts. I cannot see the springs just down the hill where the Donner party stopped on its fateful way to California, although Peter Hovingh, who knows the Great Basin from mountain to floor, assures me they are there.

Ice is everywhere. A glaze of white, breathed over the landscape, coats the ground thatch and branches. The sun blazes behind the clouds. Then the mists lighten, outlines clear, shapes begin to gain edges. As the day brightens, colors emerge—an umber stubble in the field, silvery weathered wooden fence posts strung with platinum barbed wire, greasewood frosted silver.

Vapors steam off a series of warm springs set in a cold meadow. Frost bedazzles the plants rimming the pools and puddles with a quarter-inch covering of crystals. Even as I watch, the crisp points lose their sharpness. A prairie falcon hunts, the only movement in the landscape. In the still, steely coldness of this day, it is an act of faith to imagine this spring resonating with a chorus of spadefoot toads.

In just a couple of months Great Basin spadefoot toads will breed in these ponds and springs, setting miles of the desert alive with their hoarse chortlings after a rain, a breeding invitation that varies with each species. Their name comes from the flange on their hind feet with which they burrow into the soil, rotating their sturdy bodies while pushing first with one foot, then the other, digging out of sight in moments. Spadefoot toads belong to the genus *Scaphiopus,* different from the true *Bufo* toads, two of the separating differences being the digging spade and vertical pupils. Most spadefoot toads are keyed to temporary pools and puddles. Great Basin spadefoots, breeding in these more or less permanent springs, are the exception, responding to temperature more than rain.

I walk the edge of the steaming pond, imagining them enduring this crystalline winter. Burrowing enables them to endure a season of both cold and desiccation safely underground. While hibernating, they can store almost a third of their body weight as diluted urine to forestall water loss through their skin to the surrounding soil as the soil dries. Some species may even form impervious cocoons of mud around themselves to reduce desiccation. For nine to ten months they survive without any food, locked away in a Stygian subterranean darkness, metabolic activities reduced to a barely detectable level, mere nuggets of life.

A heavy rain with air temperatures above 52 degrees F stimulates the magic of the toads' rapid emergence, a breaking through into fresh air, even though they may continue to return to a damp burrow during the day. The incessant calling of a group of male spadefoots attains an amazing volume of sound that buckles up the night air. Male calling attracts the females—males hold onto the slippery-skinned females by their embossed front fingers—and there is an orgy of mating the likes of which hasn't been seen since ancient Rome.

Along with red-spotted toads, the spadefoot's egg and larval development is among the fastest of the amphibians. Eggs hatch in two days (a frog's take a week). In nature's nicety of timing, the nourishment of the egg is gone by the time the mouth parts of the tadpole develop, and the tadpole is able to feed on its own.

In my mind's ear I remember countless pools on countless spring backpack trips into the canyons, little wigglers sometimes scattered, sometimes clustered in dense groups. Toward the end of their water period the tadpoles often aggregate in bundles, an activity apparently thermally controlled. Such aggregation not only increases their body temperatures and their metabolism, but warms the surrounding water, increasing the growth of that upon which they feed. The presence of dead tadpoles in the pools stimulates the growth rate of those remaining. In some species, the tadpoles are both carnivorous and herbivorous, with mouthparts formed accordingly. The carnivorous ones attack and eat their siblings should food become scarce, assuring that at least some will survive, and those cannibalized contribute to the survival of the species.

Scaphiopus tadpoles usually leave the water while their tadpole tails are still attached. They slither out of the pool and complete their metamorphosis, a rapid process that prepares the little swimmer for life on land as a toad in just a few hours. This sudden change in environment demands a whole new behavior. The danger of desiccation with a large surface-to-volume ratio is great and commits them to a nocturnal existence, although they can withstand some of the highest temperatures of all desert amphibians known. On this frigid morning I think of them safely below ground, withstanding the winter buried deep in the mud. I hope their feet are warmer than mine. . . .

From *Downcanyon: A Naturalist Explores the Colorado River Through the Grand Canyon* (1995)

One late spring evening in Blacktail Canyon at Mile 120.2, red-spotted toads and tree frogs deafen the air with trills and burrs and bleats, the vocalizing of all the newly risen frogs and toads who left wake-up calls with the rain.

The marvelous acoustics of Blacktail Canyon magnify their ancient anuran love songs, gather and ricochet them between the walls, curl and pulse them up and down the canyon—warbling tremolos of hope, toad trumpets and frog fancies, slippery invitations to be translated into strings of black pearls that hatch into hapless little black commas that unfold into popping and hopping creatures, transferring genes down through time on arpeggios of song.

Canyon tree frogs sound like Civil War Gatling guns, a prolonged "brrr-rrr-up," or sometimes the "ba-a-a-a" of a lost lamb, a surprisingly assertive sound from such little bodies. Red-spotted toads trill like an old police whistle, overblown and shrill, coming in sustained bursts and at wider intervals than the tree frogs' shivarees. Neither have a specific breeding time, and both respond to rainfall and warm temperature. In this narrow canyon they rise to rock-concert decibels on warm spring nights. To use the temporary water as quickly as possible demands an intense call-forwarding communication system, one of the ways desert amphibians cope with this uncertain environment. Call individuality in males and call discrimination in females become acutely developed. If someone took the time to record and analyze individual calls, the researcher would probably find that each toad has a distinctive call to designate a pool as "home."

Adult toads and frogs have a good survival rate here but tadpoles don't, and red-spotted toads have developed a specific life strategy for this situation. Of all known toads, they spend the shortest time both in the egg and as tadpoles and can tolerate some of the highest heat loads. Most are vegetarians, tugging away at algae so vigorously their little bodies quiver with the effort, but a few have carnivorous mouthparts and consume both animal and vegetable matter. As the pools dry, these few may cannibalize their less fortunate brethren, enabling them to mature more quickly and for some, at least, to survive and pass on their genes.

In the canyon suitable pools and puddles of water are likely to be separated by long distances. Generations of toads may spend their entire lives in the same pool and undertake the hazards of migration only when their population builds beyond the capacity of the pool to support them. The more complete the islandlike isolation of populations, the less chance there is of emigration and interbreeding, and the more chance for both the rates of speciation and extinction to accelerate. The river may serve as a migration path, but for animals who raft downstream on debris it's a one-way trip.

Between the pulses of the trilling, crickets chirp doggedly, woefully outsung. For them it must be like being in a room filled with a thousand cash registers going off all at once and no way to shut them off.

TERRY TEMPEST WILLIAMS
From *Desert Quartet* (1995)
Chapter 2: "Water"

At first I think it is a small leather pouch someone has dropped along the trail. I bend down, pick it up, and only then recognize it for what it is—a frog, dead and dried. I have a leather thong in my pack which I take out and thread through the frog's mouth and out through its throat. The skin is thin, which makes a quick puncture possible. I then slide the frog to the center of the thong, tie a knot with both ends, and create a necklace, which I wear.

I grew up with frogs. My brothers and cousins hurled them against canyon walls as we hiked the trail to Rainbow Bridge when Lake Powell was rising behind Glen Canyon Dam.

I hated what they did and told them so. But my cries only encouraged them, excited them, until I became the wall they would throw frogs against. I didn't know what to do—stand still and soften their blow by trying to catch each frog in my hands like a cradle, or turn and run, hoping they would miss me altogether. I tried to believe that somehow the frogs would sail through the air in safety, landing perfectly poised on a bed of moss. But, inevitably, the tiny canyon frogs, about the size of a ripe plum, quickly became entombed in the fists of adolescents and would die on impact, hitting my body, the boys' playing field. I would turn and walk down to the creek and wash the splattered remains off of me. I would enter the water, sit down in the current, and release the frog bodies downstream with my tears.

I never forgave.

Years later, my impulse to bathe with frogs is still the same. Havasu. It is only an hour or so past dawn. The creek is cold and clear. I take off my skin of clothes and leave them on the bank. I shiver. How long has it been since I have allowed myself to lie on my back and float? The dried frog floats with me. A slight tug around my neck makes me believe it is still alive, swimming in the current. Travertine terraces spill over with turquoise water and we are held in place by a liquid hand that cools and calms the desert.

I dissolve. I am water. Only my face is exposed like an apparition over ripples. Playing with water. Do I dare? My legs open. The rushing water turns my body and touches me with a fast finger that does not tire. I receive without apology. Time. Nothing to rush, only to feel. I feel time in me. It is endless pleasure in the current. No control. No thought. Simply, here. My left hand reaches for the frog dangling from my neck, floating above my belly and I hold it between my breasts like a withered heart, beating inside me, inside the river. We are moving downstream. Water. Water music. Blue notes, white notes, my body mixes with the body of water like jazz, the currents like jazz. I too am free to improvise.

I grip stones in shallow water. There is moss behind my fingernails.

I leave the creek and walk up to my clothes. I am already dry. My skirt and blouse slip on effortlessly. I twist my hair and secure it with a stick. The frog is still with me. Do I imagine beads of turquoise have replaced the sunken and hollow eyes?

We walk. Canyons within canyons. The sun threatens to annihilate me. I recall all the oven doors I have opened to a blast of heat that burned my face. My eyes narrow. Each turn takes us deeper inside the Grand Canyon, my frog and I.

We are witnesses to this opening of time, vertical and horizontal at once. Between these crossbars of geology is a silent sermon on how the world was formed. Seas advanced and retreated. Dunes now stand in stone. Volcanoes erupted and lava has cooled. Garnets shimmer and separate schist from granite. It is sculptured time to be touched, even tasted, our mineral content preserved in the desert.

This is the Rio Colorado.

We are water. We are swept away. Desire begins in wetness. My fingers curl around this little frog. Like me, it was born out of longing, wet, not dry. We can always return to our place of origin. Water. Water music. We are baptized by immersion, nothing less can replenish or restore our capacity to love. It is endless if we believe in water.

We are approaching a cliff. Red monkey flowers bloom. White-throated swifts and violet-green swallows crisscross above. My throat is parched. There is a large pool below. My fear of heights is overcome by my desire to merge. I dive into the water, deeper and deeper, my eyes open, and I see a slender passageway. I wonder if I have enough breath to venture down. I take the risk and swim through the limestone corridor where the water is milky and I can barely focus through the shimmering sediments of sand until it opens into a clear, green room. The frog fetish floats to the surface. I rise too and grab a few breaths held in the top story of this strange cavern. I bump my head on the jagged ceiling. The green room turns red, red, my own blood, my own heart beating, my fingers touch the crown of my head and streak the wall.

Down. I sink back into the current, which carries me out of the underwater maze to the pool. I rise once again, feeling a scream inside me surfacing as I do scream, breathe, tread water, get my bearings. The outside world is green is blue is red is hot, so hot. I swim to a limestone ledge, climb out and lie on my stomach, breathing. The rock is steaming. The frog is under me. Beating. Heart beating. I am dry. I long to be wet. I am bleeding. Back on my knees, I immerse my head in the pool once more to ease the cut and look below. Half in. Half out. Amphibious. I am drawn to both Earth and water. The frog breaks free from the leather thong. I try to grab its body but miss and watch it slowly spiral into the depths.

Before leaving, I drink from a nearby stream and hold a mouthful—I hear frogs, a chorus of frogs, their voices rising like bubbles from what seems to be the green room. Muddled at first, they become clear. I run back to the edge of the pool and listen—throwing back my head, I burst into laughter spraying myself with water.

It is rain.

It is frogs.

It is hearts breaking against the bodies of those we love.

Part V

Reading the Signs of the Times
Declines, Deformities, and Biodiversity

At first the signs prompted only isolated anecdotes. Herpetologists from around the globe had been noticing unusual deaths and declines among some of their chosen research subjects since the late 1970s, quietly wondering or whispering about what might be going wrong, even fearing that they themselves might somehow be at fault. But at the First World Congress of Herpetology, held in Canterbury, England, in September 1989, the numerous independent accounts coalesced and soon became amplified into the herpetological story of the decade. In February 1990, a group of researchers, led by David Wake of the University of California, Berkeley, met in Irvine, California, expressly to compare notes about disappearing amphibians and to decide on possible courses of action. This group's central conclusions and recommendations are briefly summarized in the second selection in this part, titled "Amphibians as Harbingers of Decay."

Late in 1990, Wake formed the Declining Amphibian Populations Task Force (DAPTF), which has grown into an extensive international network of professional researchers and volunteers operating under the auspices of the Species Survival Commission of the World Conservation Union. The DAPTF is also linked to the Biodiversity Programs Office of the Smithsonian Institution and is headed by Timothy R. Halliday and W. Ronald Heyer, whose article here surveys the evolution of research on the problem and considers various causes. The causes indeed seem to be multiple. One of the interesting new suspects, which was brought to light fairly early on by Oregon ecologist Andrew R. Blaustein, is the increasing level of ultraviolet radiation reaching the earth's surface as a result of the thinning of the planet's ozone layer. Other possible or probable causes of the declines of given populations include habi-

Figure 13
Fire-bellied toads *(Bombinator igneus)*. (From *The Riverside Natural History,* 1888, p. 328.)

tat destruction—the prime suspect—as well as pollutants such as industrial and agricultural chemicals and pesticides. The introduction of non-native species of fish and amphibians, especially bullfrogs, which then prey on native inhabitants, is another important factor. But as Stephen Leahy relates, even many bullfrogs in their native habitats are in danger. The most recent evidence points to a parasitic fungus that breeds on a frog's thin skin as the critical factor in the demise of at least some populations, as Virginia Morell's article explains. The continuing human desire to feed on frogs and to overharvest them for dissection are also elements of the problem.

In 1995, a new wrinkle was added to the global concern about amphibians. After ecological alarms sounded when a group of Minnesota middle school students on a field trip discovered numerous misshapen leopard frogs, researchers began noticing a range of deformities not only throughout the upper Midwest but also in many other states and some foreign countries. As with the declines, various causes—from pollution to a trematode parasite— seem to be at work in various places, and as Jocelyn Kaiser reports, many researchers doubt that any single smoking gun will be found to explain this multifaceted mystery.

A common refrain of the herpetologists at the forefront of these investigations is the necessity for more research. One of the overarching needs is for more and better information on long-term population trends among amphibians, which would help scientists decide just how severe the recent declines really are. The goal of projects such as the North American Amphibian Monitoring Program (NAAMP), a component of the international DAPTF, and Frogwatch USA, which is coordinated by the U.S. Geological Survey Patuxent Wildlife Research Center, is to collect such data through the collaboration of both professionals and volunteers. Similarly, the North American Reporting Center for Amphibian Malformations (NARCAM) was established to collate information regarding deformities. Although amphibians would probably be considered "enigmatic microfauna" rather than "charismatic megafauna" like wolves and grizzly bears, frogophiles abound, and when these creatures appear to be in trouble, even the federal government can become concerned. Thus, in February 1999 Bruce Babbitt, secretary of the Department of the Interior, announced the formation of the Taskforce on Amphibian Declines and Deformities (TADD) to fund and coordinate investigations of amphibian populations on federal lands across the country.

The declines and deformities have caused much concern in both the scientific community and the general public largely because many ecologists consider amphibians to be indicator species. The orbit of amphibian life encompasses both water and land and includes both herbivorous and carnivorous habits. Most species begin in the midst of vulnerable, unshelled eggs and spend most of their lives with the mixed blessing of thin, permeable skin. And they tend to stay put, moving about in fairly limited territories, usually rather humble territories that they often share with humans. For all these reasons, their health is perhaps an especially sensitive indicator of the health of the larger ecosystems in which they live—they are canaries in the planetary coal mine. Some researchers argue that for amphibians to be truly useful bioindicators, much more exact correlations between physiological disorder and environmental degradation would be required. And it is interesting to note that although we call them early warning systems, the warning is only early in relation to us, or at least in relation to organisms other than the mortally affected amphibians. But, of course, anthropocentric arguments for preserving biodiversity are the most appealing ones to the greatest number of people, and thus they perhaps hold the greatest promise for garnering additional research support.

Confronted with the plight of amphibians, many scientists have expanded their concern for these creatures beyond traditional questions of zoology to the difficult challenges of conservation, a development Kathryn Phillips describes in the wide-ranging introductory piece from her *Tracking the Vanishing Frogs*. This is certainly a hopeful sign because the most successful environmental movements have always combined science and activism. However, looking back over the long tradition of natural history, or of biological science in particular, one must wonder whether environmental arguments or activism based solely on science will ever be enough to move most people to adopt lifestyles conducive to preserving biodiversity. Will the arguments for saving amphibians need to make the point that we share not only an ecosystem but also, at least in some small degree, an essence? The final selection in this part is a historically grounded reflection on this possibility.

Kathryn Phillips

From *Tracking the Vanishing Frogs: An Ecological Mystery* (1994)
Chapter 1: "Lazarus"

Mark Jennings spent a blissful long weekend in the summer of 1960 camping beside the Kern River and collecting any live animal he could catch in his four-year-old hands. Mostly that meant a lot of suckers, a bony fish cast up on the shore by the fishermen who wanted something more elegant on their lines. Mark put the suckers in a bucket and tried to keep them alive. They inevitably died under his care. He also now and then spotted frogs basking on the river's banks, but his hands and reflexes and coordination weren't quick enough to nab them as they hopped into the river for cover, and so he coveted them.

The Kern River was one of young Jennings's favorite spots. It was surrounded by wild creatures that fascinated him. The Kern follows the path of one of the longest faults in the Sierra Nevada, the mountain backbone of California. Most of the river's 164 miles wind through a national park and a national forest, protected on either side by the steep slopes of the Sierra. At its southern end it arrives at the flat, dry San Joaquin Valley, with its factory-style farming and boom-and-bust oil towns.

The Jennings family camped that summer weekend along the river inside the national forest, not far from an abandoned sawmill. To get there, Roy and Helen Jennings packed Mark and his two older sisters into their car and drove 150 miles from their Southern California home, through orange groves and farmlands and vast open spaces. Then they watched the landscape turn from

From *Tracking the Vanishing Frogs* (1994)   165

flatlands to hills to distant granite peaks, from chaparral to oaks and cottonwoods to pine forests.

This was a time before *ozone* was an adjective for *hole,* or *greenhouse* a shorthand for an environmental malady. Life in the campground seemed untroubled, and being outdoors in the forest felt healthy and comforting. The family spent days fishing, relaxing, playing by the river. At night they slept in sleeping bags on the floor of a big green canvas tent that was so heavy and hot that under the afternoon sun it felt like a sauna.

On the last day of their vacation, the family folded up the tent and stowed the aluminum ice chest from J. C. Penney in the car. Mark helped where he could, but mostly stayed close to camp and waited to depart from a place he didn't much want to leave. When almost everything was packed, a boy Mark had befriended during the weekend came to the campsite to say good-bye. He carried one of the coveted frogs. Mark was impressed. Then the boy presented the frog to him as a going-away gift. Delighted, Mark carefully took the creature in his hands. Its back was bumpy and olive-colored and had some spots and mottling. It had a triangle of paler olive on its snout. Its legs were a bright yellow on their undersides. It was perfect.

Mark's parents, both schoolteachers who cheerfully tolerated their son's fascination with furless animals, found an empty mayonnaise jar to hold the gift. They put a little water in the jar. Then Mark held the occupied container on his lap on the long ride home.

The family deviated from their normal route home and went east over the mountains into the Mojave Desert for a change of scenery before they headed south and west. Scrub and creosote bushes and miles of sand and dirt that seemed to change color with the light lined the roads. The car had no air conditioning and the ride was hot. The hours were made hotter for the frog by the curious boy who held it in the jar. Occasionally, Mark would shake the jar to make sure the frog was still alive. The frog moved less as the hours passed. The shaking and the heat proved too much for the creature.

Roy Jennings pulled the car into the family's driveway in Santa Paula, a pinpoint on the map about an hour's drive northwest of downtown Los Angeles, and checked the frog's condition. It was still and limp. While Mark's mother tried to console the boy, his father opened the jar and dumped the poor frog's body into a patch of ivy beside the driveway.

Later that evening, Roy Jennings put the family dog on its leash and took it for a walk. As he came back up the driveway, he saw movement in the ivy. It was the frog, hopping and miraculously alive. The elder Jennings captured it and took the creature inside to his son. The next day, father and son went to a pet store and bought a five-gallon aquarium to become a home for the frog. They put gravel and some water on the bottom. Then Roy Jennings fashioned a screen for the top. He used brass screws and made the screen so sturdy that Mark still has it more than thirty years later. Helen Jennings made a net using an old nylon stocking that she sewed onto a hoop fashioned from a wire coat hanger. With that, Mark could catch insects to feed the frog. Helen also named it Lazarus after the biblical character whom Jesus raised from the dead.

Lazarus thrived in his glass home, which sat on a kitchen counter in the Jennings home. Mark regularly caught the insects the animal needed to have. Then one day, about a year after the frog joined the household, Mark noticed a moth weakly flapping its wings. It was so docile that he was able to catch it with his bare hands and promptly give it to Lazarus.

Frogs are notorious for snapping their sticky tongues out in a flash and grabbing virtually anything that moves. They are not picky eaters and have been known to kill themselves by overloading on gravel during a feeding frenzy. Lazarus needed no coaxing to eat the moth. Unfortunately, it proved to be his last meal. The frog was dead within hours. In retrospect, Jennings figures the moth must have been so easy to catch because it was dying from pesticide poisoning.

As I hear this story of Lazarus, I am sitting beside Mark Jennings in his white Ford Ranger pickup truck while he drives it along one of the highways near his hometown. We are headed to see if we can spot a certain rare toad along a stream in the hills not far from his parents' house. A steady but light rain falls, even though it is early July, the driest season in Southern California. It is the kind of weather that typically draws the toads and frogs from their damp hiding places as though in celebration of water and everything it means to them. I have heard parts of the Lazarus frog story before, but this time, as I listen to it from beginning to end, it seems almost prescient.

Thirty-two years have passed since the Kern River camping trip when Mark held his first frog, and many things have changed. Jennings is now six feet tall and sports a light brown mustache. His parents still live in the same Santa Paula house, but he moved long ago and now lives alone in a one-bedroom apartment that doubles as his laboratory and office in Davis, in Northern California. He continues to like frogs, but now his interest is scientific as well as aesthetic. He is a biologist. His doctorate is in ichthyology—the study of fish—but his interest is primarily in amphibians—frogs and toads, salamanders and newts, and caecilians. Though he once thought frogs were everywhere and forever, as an adult Jennings has become a chronicler of their demise and disappearance.

The chronicling began almost unconsciously with Lazarus and his kind. Lazarus was a foothill yellow-legged frog, whose scientific name is *Rana boylii*, one of the five frogs of the genus *Rana* that are native to California. Ranid frogs are often called "true frogs" and are represented by various species in practically every part of the world. More than thirty species of ranids are found in the United States alone. Among these is the northern leopard frog, *Rana pipiens,* a creature often used in high school biology classes of the past. Another is the common bullfrog, *Rana catesbeiana,* the critter whose legs are served in certain restaurants.

Ranids look like what many people, especially Americans and Europeans, think of when they think of frogs. They have long hind legs, long toes with webbing on the rear feet, and their topsides are green or brown or some variation of either color. When fully grown, most average a size that fits neatly into the palm of an adult human's hand, but they can get even larger.

Foothill yellow-legged frogs like to live in riffling streams with rocky or cobblestone bottoms loaded with crannies in which they can hide to escape predators. They like it where the banks and water surfaces are covered by a mix of sun and shade, the same sort of idyllic setting that a weekend fisherman wants. The Kern River has lots of spots like that, and it is still popular for camping and fishing. But if Jennings camped there today, he wouldn't find another Lazarus. There have been no confirmed reports of any sightings of foothill yellow-legged frogs on the Kern River since the early 1970s. In fact, Jennings says, that frog is gone from 45 percent of the places it once lived in California.

When Mark was a boy, foothill yellow-leggeds could be found in streams in much of Northern California and part of Southern California. Jennings recalls one time he saw a foothill yellow-legged in the east fork of Santa Paula Creek in the hills above his hometown. It was May of 1970. "We have the date on Mom's little calendar," he explains. He was on a Boy Scout camping trip, and his mother routinely noted such occasions in calendars that she then saved after the year was out. Jennings returned to that stream and others nearby many times over the succeeding years. But that one trip stands out: "That was the last time I saw *Rana boylii* in Southern California."

Under normal circumstances, Jennings's story about the foothill yellow-legged frog might be interesting, but no cause for great concern. Papers noting an apparent disappearance or decline of some species of frog or salamander occasionally have surfaced throughout the history and literature of herpetology, the science that studies amphibians and reptiles. But since the late 1970s, and particularly since the early 1980s, stories recounting unquestionable declines like that of *Rana boylii* have become so common that scientists are alarmed.

In Australia, for instance, scientists discovered in 1974 a rare frog that actually regurgitated its babies. The mother frog's odd physiology held promise for use as a model for devising better treatments for stomach ulcers. Six years later the two known populations, harbored in an apparently pristine rain forest, disappeared. The frog hasn't been seen since. Researchers estimate that twenty-six of Australia's 202 species of frog are in trouble.

In Costa Rica, photos of the orange-colored golden toad, *Bufo periglenes,* have commonly been used on tourism posters promoting the country's wild attractions. As late as the summer of 1992, frog lovers were traveling to the mountain village of Monteverde, on the edge of the toad's protected forest home, in hopes of glimpsing it. But the toad hasn't been seen in significant numbers since 1987, and not at all since 1989. It disappeared almost overnight. At the same time, several other varieties of frogs and toads and salamanders disappeared or dramatically declined in the Monteverde area.

In the mountains of Oregon, Colorado, and California, several kinds of amphibians, including Yosemite toads, Cascades frogs, leopard frogs, and western toads, dropped in number in the early 1980s and have not recovered. Some of these animals were once so abundant that scientists and backpackers recall having to take care not to step on them when hiking in the high country. One informal survey of scientists in the northeastern United States suggests there may be declines there in as many as twelve types of salamanders and eight types of frogs.

Scientists report having a difficult or impossible time finding certain species of once-abundant frogs in the Peruvian and Ecuadoran Andes. In Puerto Rico, one type of common frog has not been seen in a decade, even though there has been no obvious change made to its habitat.

The list of declines became so long and unnerving by the late 1980s that scientists who work with frogs slowly and cautiously began sharing their tales of frog-finding difficulty with their colleagues. Among herpetologists, particularly those who work in the field more than in laboratories, being able to find amphibians traditionally has been a matter of pride. If a graduate student returned from a summer field trip complaining about not being able to find animals, colleagues and teachers assumed the student was not very competent. So when the professionals began sharing their experiences of not finding frogs, it was almost cathartic, like releasing some horrible family secret.

The notion that something as common as a frog should be disappearing is a strange one, considering amphibian history. Scientists estimate that the first amphibians appeared about 350 million years ago. At the time, the planet Earth contained one huge landmass and one even larger ocean. Plants and insects lived on land, and vertebrates—animals that have backbones and nervous systems—in water. Among the vertebrates was a class of bony fish called *Osteichthyes,* and within that class there were, among others, lungfish and lobe-fin fish. Somehow, one of these two fish—which one is a matter of scientific debate—evolved into a creature that had legs where fins would normally be and a physiology that allowed it to leave the water and walk on land for long periods. This creature crawled out of some swamp somewhere and a new class was born: Amphibia. Animals in this class were mostly terrestrial but usually lived in water at some point during their life cycles.

Over the eons, amphibians have evolved from a single group of large animals into three groups or orders of small animals: Caudata, Anura, and Gymnophiona. The order Caudata includes salamanders and newts, the slender, four-legged, tailed animals that look something like lizards with scaleless bodies. There are about 390 species of Caudata, and most are found in the Northern Hemisphere.

Gymnophiona are caecilians, secretive creatures that live under leaf cover or underground. They look more like very large worms than anything else. Scientists know relatively little about caecilians, and the number of people in the world who study them could probably fit into a standard-size hot tub. There are about 160 known species of caecilians.

The third order, Anura, are frogs and toads. Anurans are the most abundant amphibians and are represented by about 3,960 species. They also are the noisiest amphibians, and the croaking or whistling call of males looking for mates is one of the harbingers of spring in many parts of the world. The distinction between frogs and toads is imprecise. In general, toads are wartier, have poison glands, and can tolerate drier conditions. Toads are types of frogs, but frogs are not types of toads. Herpetologists often use the word *frog* to refer to both.

Scientists estimate that the first frog evolved from the main amphibian lineage about 150 million to 200 million years ago. Researchers usually date the beginning of complex life on Earth at 600 million years ago. Within those 600 million years, at least five—and some scientists say six—major global mass extinctions stripped the world of many of its creatures. At least two of the mass extinctions occurred after the appearance of frogs. The most famous was the one 65 million years ago that some researchers attribute to the effects of an asteroid's collision with the Earth. The collision churned up so much dust, according to one theory, that the Earth was darkened, precipitating climatic changes and food shortages. The great dinosaurs disappeared during this period. But the frogs survived. Scientists estimate that during each mass extinction at least half of all animal species died. Yet through at least two of these events, frogs have endured.

But something is very different in the world today from the way it was 150 million or even 65 million years ago. Humans have evolved and become, in very quick order, the Earth's dominant and most destructive animal. Formerly, a set of naturally occurring events would prompt earthly environmental cataclysms like mass extinctions. No single creature could be blamed for these events; they just happened from a combination of geologic and atmospheric conditions. Now, though, environmental disasters are routinely created by humans, and real cataclysms are within our ability to cause as well. People like Mark Jennings worry that significant numbers of amphibian species are facing another mass extinction. This one, though, seems to be brought about by humans at a breakneck pace. And this one may be the one that the frogs fail to survive, unless someone does something soon to stop it.

Other animals that share the late twentieth century with humans and frogs are also in danger. Large mammals—bears, wolves, and the like—have declined dramatically as their habitats have been wiped out. Birds have suffered similarly. Any amateur naturalist can name at least one bird that has captured headlines because its numbers seemed to tilt toward extinction. Peregrine falcons, brown pelicans, bald eagles, California condors, and a list of songbirds are just a few, not to mention the famous spotted owl of the Northwestern old-growth forests.

Most people know about the plights of these feathered and furred animals because they have had strong advocates. They are, as wildlife managers sometimes call them, "charismatic megafauna." They are animals who touch a public nerve, draw attention, and get action. They are represented by groups such as the Audubon Society, first organized in the late nineteenth century to stop the wholesale slaughter of birds for their feathers, or the National Wildlife Federation, originally started to represent the interests of the gun-and-rod crowd, who wanted to make sure there would be plenty of wild land and wildlife around for recreational enjoyment. Even state and federal fish and game agencies have a history of advocating the preservation and protection of these megafauna.

Frogs and toads, on the other hand, haven't had such advocates. They are hairless, slick, sometimes slimy, and even warty—characteristics that many people find aesthetically displeasing and even offensive. Most of them don't have any value as edible game; eating some would send you to your grave. You can't hug a frog. Unless you have a very unusual sense of style, you can't easily mount a dead frog and decorate a room with it—although in some parts of the world you can buy a small purse made from a giant toad's skin. But even that item seems more tasteless than valuable. So the frogs and toads, despite a long history of important roles in art, literature, folklore, and fertility rites, have arrived at the end of this century in dire straits.

Amphibians are facing what some herpetologists consider a crisis. Some even suggest that the frog declines may be indicators of greater environmental problems. By 1990, a decade or more after the crisis began for many of the species, a collection of scientists raised the first loud alarm that something was wrong. Then the scientists took on the challenge of responding to the crisis.

Different scientists responded in different ways. Some formed committees to oversee more committees to figure out how to respond, much like any bureaucracy. Some headed back out to marshes and ponds to actually count frogs and examine their living conditions in an effort to identify exactly which amphibians were declining and which ones were doing fine, and why. Some drafted grant proposals and received funding for multiyear studies that included lab experiments to answer some of the toughest unanswered questions about declines. Some worked harder to protect the amphibians they knew were in danger. One, for instance, spent many months battling motorcycle gangs, more politely called off-road vehicle enthusiasts, who routinely plowed their noisy vehicles through a rare toad's mating grounds with the U.S. Forest Service's tacit approval. Some did a little bit of everything.

The way each scientist has acted on the declining frog issue has usually reflected individual interests and philosophies. Some are mostly concerned about how the research should be conducted. Others believe frogs could be a good vehicle for making other environmental and scientific issues more compelling. Some, like Mark Jennings, are simply passionately interested in frogs, particularly rare frogs. The single thread tying all these scientists together has been a desire to know why frogs are disappearing.

Sometimes their task is like piecing a mystery together. At other times it is simply acknowledging and then quantifying what seems perfectly obvious. All the while, some of the researchers worry that they are witnessing a global environmental disaster that merely includes frogs as early victims and could end with humans. If such a disaster is in the works, will the typically slow pace of science mean that they will identify the disaster's cause only after it is too late to be corrected?

Chapter 9: "A Place for Frogs"

Shortly after publication of biologist George Schaller's book about efforts to save the diminishing panda population in China, Schaller gave voice to a growing sentiment among scientists who study wildlife. "Research is fun and it's easy. But no scientist can afford just to study," he told *The New York Times* in a brief interview. "There's a moral obligation to do more for conservation. If you only study, you might get to write a beautiful obituary but you're not helping to perpetuate the species."

The group of scientists who joined in Irvine in 1990 to notify the world about the declining amphibian problem did so believing something needed to be done, that time was running out for some frogs, toads, salamanders, newts, and caecilians. Also, they knew that a variety of environmental problems were smoldering around the world, and examining amphibian declines provided a focus for examining the impacts of those problems. They believed that science and scientists could be tools to get something done. For most of those involved in the issue, being a useful tool has meant doing the normal stuff of science: conducting research, collecting data, doing experiments, writing papers.

In the fall of 1991, [Mark] Jennings and [Marc] Hayes decided they needed to do more than scientific work to save amphibians. They knew the

California red-legged needed protection. Its best chance for protection, they figured, was to get it included on the federal list of endangered species under the Endangered Species Act. But to get on the list, someone had to nominate the frog. The Fish and Wildlife Service could nominate it, but years could pass before the paperwork finished winding its way through the understaffed agency's hierarchy. Outside parties could also nominate the frog. An outside nomination would likely result in a listing sooner, because the act gave the agency specific, swift deadlines by which it had to make a decision on outside nominations.

Nominating an animal involves more than just submitting a species' name. The nominator has to provide a detailed account of the evidence that the animal needs protection. That evidence must be backed by published and unpublished scientific research. The Fish and Wildlife Service has been known to reject deserving animals because a nomination was poorly compiled and unconvincing.

Jennings and Hayes knew that getting involved in the nominating process would be time-consuming. They already had many projects sitting on hold while they finished their statewide survey of amphibians and reptiles. They weren't anxious to put even that survey on hold to compose a nomination. At the same time, though, they were worried that a delay of several years could be disastrous for some California red-legged populations. So, after a few months of discussion and writing, "the two Marcks" completed a nomination for the California red-legged frog and submitted it to the Fish and Wildlife Service in January 1992. It was more than eight single-spaced pages long, not including its maps and bibliography of supporting scientific literature.

"Evidence for the disappearance of the California red-legged frog is most consistent with four types of human interference . . . (1) loss of habitat; (2) fragmentation of habitat to produce deleterious area and demographic effects; (3) overexploitation; and (4) the spread of exotic (introduced) species," Jennings and Hayes wrote in the nomination. "A fifth major class of human interference, pollution, may have contributed; and a sixth, climate change, has a significant possibility of detrimentally affecting this taxon in the near future. . . .

"The California red-legged frog . . . is estimated to have disappeared from over 99 percent of the inland and southern California localities within its historic range and at least 75 percent of all localities within its entire historic range," the two concluded. "Many of the factors believed responsible for the extirpation of [the red-legged] from localities within its historic range still affect the large majority of populations that remain, and future conditions are anticipated to become even less favorable for this taxon."

The Endangered Species Act has been under attack almost since the day it was adopted into law, in 1973. The act relies primarily on biological evidence to determine whether an animal is eligible for listing as either threatened or endangered. Depending on the level of its listing, any activity that would hurt the listed animal—from destroying habitat to owning or selling the animal—is prohibited. The act's principal critics are developers, logging companies, utilities, and other commercial interests who complain that complying with the act costs too much money, time, and jobs. The Secretary of the Interior can, in fact, waive the act for a species if it is causing "undue economic hardship."

In reality, the act has been neither as damaging as its critics claim nor as effective as its supporters would like. Few development projects have been stopped. Logging has continued in most forests, despite the presence of declining species. Many animals that deserve listing have died without notice.

The act has rarely been backed up with the kind of government funding and other support that would be required to implement it fully. The Fish and Wildlife Service has never had as many staff members as it needed to compile and process the complex nominations as quickly as the law requires, or to put into effect programs to help endangered animals recover from their precarious situations. During the 1980s, under the antienvironment White House regimes of Ronald Reagan and George Bush, government support for the endangered species program dipped to an all-time low.

A 1990 audit by the U.S. Department of the Interior's Inspector General found that six hundred candidates that the Fish and Wildlife Service deemed eligible for listing had not been listed. In addition, the service had identified another three thousand species that it suspected of being eligible, but on which it still had taken no action. "During the last ten years," the audit said, "at least thirty-four animal and plant species have been determined to be extinct, without ever having received full benefit of the act's protection, and those species currently known to merit protection, as well as those candidate species eventually determined to need protection, are similarly in jeopardy of extinction."

When Jennings and Hayes decided to nominate the California red-legged for listing, they expected the process to move fairly quickly. They knew their nomination package was strong and backed by solid research. Jennings, who was in California and routinely in contact with the U.S. Fish and Wildlife Service offices, took on the task of being primary contact on the nomination. He was afraid the application would get lost in the paperwork maze if he didn't keep tabs on it.

Even then, though, things moved more slowly than he expected. The act requires the Fish and Wildlife Service to decide within ninety days whether there is enough information to consider listing a nominated animal. It took the agency more than nine months to make its decision. The act also requires the agency then to decide within one year of the nomination whether to list an animal. Less than a month before that one-year deadline was up, Jennings got a call from a harried Fish and Wildlife Service biologist who had just been given the job of reviewing the application.

Jennings and Hayes had attached the red-legged nomination to a nomination for listing of the western pond turtle, which shared the same riparian habitat. The turtle's nomination, cosponsored by another scientist, was riddled with problems and had helped slow the red-legged paperwork, Jennings learned. Now the two had been separated, and the red-legged was getting closer attention. And the agency biologist needed copies of all the papers about the red-legged that were referred to in the nomination. Could Jennings help? For the next two days, Jennings stood in a photocopy shop, putting a boxload of papers together. "My feeling is that if I propose something, I'm going all the way," he said later. "If you want the animal listed, you have to be available to provide the data and stand by it."

Meanwhile, Jennings and Sam Sweet decided to join forces to get the arroyo toad listed. The toad had earlier gotten entangled in an internal Fish and Wildlife Service dispute that centered on the interpretation of part of the Endangered Species Act. As a result, the service's own nomination of the toad hit a dead end. Sweet and Jennings decided to avoid more wasted time by nominating the toad themselves.

Finally, as the Fish and Wildlife Service missed deadlines to act on each of the nominations, the Environmental Defense Center, an environmental law firm in Santa Barbara, got into the picture. The firm notified the agency that it intended to sue unless the agency hurried up and made its decisions on the animals. The threat of a lawsuit has become a common tactic to push the Fish and Wildlife Service to respond to listing nominations. The sad truth, as one of the agency's biologists told me, is that the agency is so swamped that it chooses which Endangered Species Act nominations to respond to first based on which ones have lawsuits attached.

Late in July 1993, I got a phone call from a very cheerful Jennings. He had just heard that the Fish and Wildlife Service had published in the *Federal Register* its intention to list the California red-legged frog. "That means that the major hurdle is over," Jennings explained. The notice didn't say whether the frog would be listed as threatened or endangered—two levels that carry different legal weight—but Jennings didn't mind. Either way, the California red-legged would become the first North American frog on the federal endangered species list. "The goal was to get it listed. What was critical was to get it listed so they keep from sliding into oblivion."

But already he was anticipating opposition to the listing. It could result in more protection of riparian habitat, the bushy streamside ecosystem that is disappearing about as fast as the red-legged. Right away, two proposed State

Water Project dams that Jennings said would "blow out" red-legged populations on the "last good streams" in the hills on the west side of the San Joaquin Valley could be affected by the listing.

Three years earlier, when I'd first joined Jennings on a field trip to see California red-leggeds, he had been decidedly pessimistic about their future. Now, finally, he has reason to be optimistic. "Does it help the frog survive into the future?" he asks rhetorically. "Oh, yes it does!"

Laurie J. Vitt, Janalee P. Caldwell, Henry M. Wilbur, and David C. Smith

"Amphibians as Harbingers of Decay" (1990)

Evidence for declines in amphibian populations was discussed by researchers from around the world at a conference in Irvine, California [February 19, 1990]. Three conclusions emerged from the meeting, which was sponsored by the National Research Council's Board on Biology. First, certain amphibians, even in habitats that appear to be pristine, are disappearing at an alarming rate, and the declines are widespread and have been particularly serious since the late 1970s. Second, among the vertebrates, amphibians may be the best animals to use as biological indicators of ongoing and impending environmental degradation. Third, the data necessary to ascertain changes in amphibian populations, and to determine the underlying causes, are virtually nonexistent.

Why should we worry about amphibian declines in remote areas, and what is it that makes amphibians good bioindicators? Amphibians are certainly central to many ecosystems, with their high position in the food chain and a biomass in many habitats that often exceeds that of all other vertebrates combined. In addition, amphibians have complex life histories, usually with aquatic eggs and larvae and terrestrial adults. The adults of many species spend their lives against or in the substrate (e.g., mud, sand, or leaf litter) and water. The skin of amphibians at all life-history stages is permeable to water and many electrolytes. The eggs are covered only by a layer of gelatinous material, so they are directly exposed to the environment. Research has already revealed that embryos, as the cells cleave, are sensitive to pH and heavy metals. In addition, an assortment of anions and cations may influence mortality at the various life-history stages. The disappearance of amphibians in remote areas suggests a general degradation of the environment, possibly worldwide in scope.

What should be done? Central to the recommendations of the conference attendees is the establishment of strategically placed monitoring programs that take into account geographic, taxonomic, and habitat diversity. The monitoring would, by necessity, include long-term population studies, measurement of physical and chemical environmental characteristics, and appropriate experimental studies dictated by individual systems.

And herein lies the problem. Funding agencies are often wooed by the esoteric demands of high-technology science. Research on environmental degradation and on the impending loss of biodiversity is woefully underfunded. A recent report by a National Science Foundation task force on the biodiversity crisis notes that one-quarter to one-half of the earth's species will become extinct in the next 30 years (*Science,* 16 February 1990). Under current priorities, even the prospect of a thin web of biomonitoring research stations seems beyond reach. Ecologists, systematists, and funding agencies are long overdue to adjust their perceptions and respond to the urgent need for global inventories. As already called for by many prominent conservation biologists, we recommend the immediate marshalling of a major biodiversity project, on the scale of the Human Genome Project or even larger. The survival of our planet as we now know it lies in the balance.

Andrew R. Blaustein

"Amphibians in a Bad Light" (1994)

Each spring, two or three times a week, my graduate students and I travel eighty miles east, from Oregon State University to Lost Lake in the Cascade Range. Surrounded by snow-capped peaks and volcanic debris, the lake, at an altitude of about 4,000 feet, is the lowest in the area and consequently the first to lose its ice covering. And when the ice goes, western toads *(Bufo boreas)* immediately begin breeding by the hundreds. Because this can happen any time in May or June, frequent visits insure that we catch them in the act. But it is not easy; some years, when the snowpack is great, we have to snowshoe several miles around the lake to get to the breeding site.

This year, however, the snow around the lake melted by early May, so we could drive close to the small section of the lake that these amphibians seem to prefer. Toads began to emerge on May 6, the earliest date we had ever seen them arise from their six-month winter sleep. We donned our waders, picked up our pails, nets, scales, and notebooks, and ventured into the cold, clear water. After some searching, we found hundreds of toads.

To obtain a long-term record of their reproductive patterns and changes in their population size, we had been weighing, measuring, and marking western toads at Lost Lake since 1979. Nevertheless, we were still amazed at how the toads, in their quest to mate, scrambled over snow and ice and into near-freezing water. Close to the surface, where the toads congregated, violent winds and snow, accompanied by pelting hail, continued to buffet them while they bred.

Males arrived first, followed by females, and in a seemingly haphazard manner, hundreds of toads searched for mates. Because the males outnumbered females, the competition for access to females was intense and shoving matches often erupted. Soon after the pairs formed, the females began laying eggs and the males fertilized them by releasing sperm directly into the water. On average, each female produced some 12,000 eggs in long strips surrounded by a protective, jellylike covering. The individual strings stretched for more than twenty feet and often became intertwined with the shallow vegetation and with eggs deposited by other females. Three days after the toads began to converge on the lake, egg laying was complete and they disappeared into the forests to feed and fatten up before hibernating again in the early fall.

During peak years at Lost Lake, we have seen more than 500 pairs of toads, several hundred unpaired males, and several million eggs strewn about the lake, but this year we counted only 147 breeding pairs and about 100 unpaired males. Almost two million eggs were laid. As usual, the eggs, with their jet-black embryos, began to develop normally, but two days after they were laid, we began to see the same ominous pattern we had observed for the past several years at lakes throughout the Cascades. Many embryos began to turn white as they started to die in wavelike fashion from one end of the enormous egg mass to the other. Soon they became a putrid, decaying mess, attracting flies and other insects—a potential feast that lured Pacific tree frogs *(Hyla regilla)* to the site. Opportunistic garter snakes arrived next to dine on the frogs. A week after the toad eggs had been laid, only half of them were viable. After the normal developmental period of about two weeks, even fewer hatched as tiny tadpoles. And this was only a small part of a much bigger problem.

By the mid-1980s, we began to notice that the frogs and toads we had been studying were becoming more difficult to find. Some populations of the most common species, such as the Cascades *(Rana cascadae)* and red-legged *(R. aurora)* frogs and the western toad, were nowhere to be found. Through the grapevine, I learned that amphibians were disappearing in many parts of the world, from North and South America, Asia, Africa, and Australia. Some species were even reported to have become extinct. To add to the mystery, in some areas populations of certain species were doing fine, while others were disappearing. In Oregon, for example, Pacific tree frog populations were thriving in lakes and ponds where western toad and Cascades frog populations were dwindling.

Biologists proposed many possible causes for the declines, including habitat destruction, pollution, and natural population fluctuations, but no single reason was apparent. In many areas, including our study sites, declines were occurring in relatively undisturbed habitat with no apparent pollution. Yet by the late 1980s, a pattern began to emerge that gave us some clues to the puzzling egg deaths—and perhaps to the shrinking populations—in the Cascades.

A significant number of the troubled species were mountain-dwelling amphibians that laid their eggs in the open, in shallow water—the same way in which the declining Oregon species did. Throughout development, these relatively unprotected eggs would be exposed to sunlight and potentially harmful ultraviolet (UV) radiation. The middle portion of the UV spectrum, known as UV-B, is especially dangerous. In humans it can cause sunburn and skin cancer and weaken the immune system. UV-B can also damage amphibians. In the mid-1970s, a laboratory study conducted by Robert Worrest, at Oregon State University, showed that western toad embryos developed abnormally when subjected to UV-B radiation. Since then, several reports have documented a gradual increase in UV-B radiation hitting the earth's surface as the protective ozone layer thins. Were increasing levels of UV-B radiation responsible for the high rate of egg mortality in the Cascades? And if other declining amphibians had a similar vulnerability at the egg stage, could increasing exposure to UV-B radiation each spring be responsible for shrinking their populations over time?

In the late 1980s, when I first suspected UV-B radiation caused the toad eggs to die, I began a series of simple experiments to see if this was possible. I brought some newly laid western toad eggs to my laboratory and reared them in the absence of sunlight in aquariums filled with lake water. I followed the eggs' development until they hatched. Several months later the tadpoles metamorphosed into young toads. I was excited to see that almost all of the laboratory eggs survived, while eggs from the same clutches left to develop in the lake had died at unprecedented rates. Whatever the problem was, it did not seem to be in the eggs themselves.

After hearing about my ideas on UV-B radiation and amphibian declines, John Hays, a molecular geneticist at Oregon State University, called me to discuss how he might get involved. To learn how plants and animals repair UV-B damage to DNA, Hays had been studying the process in the eggs of African clawed frogs *(Xenopus)*—the amphibian equivalent of the well-studied laboratory rat.

When a cell is exposed to UV-B radiation, the energy can be absorbed by a number of biologically important molecules, including proteins and DNA. When this happens, the bonds between atoms and molecules can be altered, causing the cell's chemical machinery to malfunction. Such changes are particularly disruptive when they occur in DNA; if the genetic code, which carries the instructions for life, is misread, mutations and cell death can occur. Many plants and animals, however, are able to repair a certain amount of DNA damage. Photolyase, an enzyme found in the cells of many organisms, can remove the harmful defects. Hays reasoned that the eggs of different amphibian species may contain different amounts of photolyase. Therefore, we predicted that species with the greatest quantities of the enzyme would be more resistant to damage by UV-B than species with less photolyase. This could explain why the eggs of some amphibians were dying while those of others were unaffected.

The first step in testing our hypothesis was to collect eggs from a number of amphibian species with different egg-laying behaviors. We also made sure to include eggs from species that were in decline and from species that were doing well. Once eggs were collected, Hays and his chief technician, Peter Hoffman, measured photolyase levels, using the same techniques they had perfected while studying African clawed frogs.

The results of the molecular tests were compelling. The eggs of the nine species we examined showed enormous differences in the amount of photolyase they contained. Eggs of the Pacific tree frog (which are laid in open, shallow water) had the most photolyase: three times as much as the Cascades frog and six times as much as the western toad. The eggs of the six salamanders we tested had less than any of the frogs, with the least amount being found in Dunn's salamander, *Plethodon dunni* (Pacific tree frog eggs had eighty times more photolyase than these amphibians).

The correlation between levels of photolyase and egg-laying behavior was striking. The salamanders, whose eggs have little photolyase, generally lay their eggs under logs, in crevices, or in deep water—all places where little UV-B radiation will penetrate—while species that lay their eggs in the open, exposed to sunlight, had the highest levels of the enzyme. The egg-laying behavior of salamanders may not have evolved specifically to afford them protection from UV-B radiation; other selective pressures, such as predation and temperature requirements for development may have been more important. Nevertheless, the protection from UV-B may be a secondary benefit. Those species that laid their eggs in the open, however, needed high levels of photolyase to minimize the damage to their DNA caused by exposure to direct sunlight.

Although the results of the enzyme studies were suggestive, we still needed to know if UV-B radiation was damaging eggs in nature. We began tackling this question even before we had the results of the DNA repair study. With field experiments in lakes and ponds where amphibians naturally lay eggs, we could compare the hatching success of eggs exposed to UV-B with that of shielded eggs. We gathered freshly laid eggs from four species that deposited them in the open: Cascades frogs, Pacific tree frogs, western toads, and northwestern salamanders *(Ambystoma gracile)*. We placed the eggs in the bottom of screened, boxlike enclosures that allowed water to flow freely through them. Over some of the enclosures, we placed plastic filters that blocked UV-B. We left a second set of enclosures uncovered, exposing the animals in them to the rays. A third set, which had a plastic filter that allowed transmission of UV-B, provided a control to insure that any variation we found under the UV-B-blocking model was not due to the presence of a plastic cover.

We placed the enclosures randomly in the shallow water of lakes or ponds where natural breeding sites were located. By using four enclosures of each type, we insured that our results were not caused by some bias in our procedure or by a small sample size. Setting up the experiments at several different sites helped assure that any results we obtained were not unique to a particular area.

Although we only had to follow the development of the eggs until they either hatched or died, the experiments took two years to complete. Like most fieldwork, the project ran into some unexpected trouble. During the first year, we could not get enough viable eggs to set up our experiments because the animals did not breed at all sites. Spring storms with high winds destroyed some of our enclosures. Under the same harsh spring weather encountered by the toads, we had to count and measure each egg, in every enclosure, every day. And then there was vandalism, both by humans and smaller mammals. So someone had to guard each site, twenty-four hours a day, until the experiments were done, which often took two weeks. By the end of the second year, however, we had results that were both dramatic and foreboding.

More than 40 percent of the western toad and Cascades frog eggs exposed to UV-B radiation died, compared with 10 to 20 percent of those that were shielded. Northwestern salamanders did not fare better; more than 90 percent of their exposed eggs died. The Pacific tree frog, however, was unscathed, with almost all of its eggs surviving under all lighting conditions.

These results, together with those from the DNA repair study, convinced us that the link between UV-B radiation and the egg deaths was real. Natural levels of UV-B were killing amphibian eggs in the field. Pacific tree frogs—which had the highest levels of photolyase and whose populations were doing fine—seemed more resistant to UV-B than did western toads and Cascades frogs, two species with less photolyase that are in decline throughout their ranges. We do not know the status of northwestern salamander populations, but given our results, they too could be in jeopardy.

We had found one small piece of the amphibian decline puzzle. But many questions remain. Is the egg mortality in the Cascades caused solely by UV-B exposure or are other factors involved? We had observed the growing presence of a pathogenic fungus in the lakes. Is the UV-B radiation compromising the defense systems of embryos, making them more susceptible to disease? Another question is, how much mortality during the egg stage can a population endure before it crashes? Western toads, for example, live for twenty years or more, so the results of their reproductive troubles in recent years may not become apparent for many years.

Finally, could we settle on a universal explanation for amphibian declines? Increasing levels of UV-B radiation are obviously not the only reason these animals were disappearing. It cannot explain, for example, why species that live under dense forest canopies, protected from UV-B, are also in trouble. Perhaps other organisms, such as plants, fish, insects, and even humans will provide us with more information on the damaging effects of increasing levels of UV-B. We know, for example, that certain crop plants have reduced growth, photosynthetic activity, and flowering when exposed to UV-B radiation. In the Antarctic, severe ozone depletion and increased UV-B have been associated with reduced growth in phytoplankton.

If projected increases in UV-B occur, over evolutionary time there may be increased selection pressure on amphibians and other organisms to evolve efficient repair mechanisms or to alter their behaviors and thereby minimize their exposure to UV-B. Unfortunately, changes wrought by human disturbance occur at such rapid rates that many organisms may not have time to adapt.

Timothy R. Halliday and W. Ronald Heyer
"The Case of the Vanishing Frogs" (1997)

In the summer of 1989, David Wake, director of the Museum of Vertebrate Zoology at the University of California at Berkeley, set out to solve a little mystery. Several months earlier he had pointed David Green, then a postdoctoral fellow, to a site in the nearby Sierra Nevada that Wake knew to be abundant in *Rana muscosa*, a mottled yellow and brown frog Green was studying because of its unusually broken distribution patterns. But when Green reached the designated location, he couldn't find a single specimen.

Puzzled by Green's account, Wake decided to accompany him to the site, assuming he had simply missed it the first time. But when they arrived, Wake, too, was surprised to find that all the adults had disappeared and only a couple of tadpoles remained.

Wake and his other students soon began to notice similar disappearances at other popular frog localities in central and northern California. Wake wondered if he had stumbled upon a bigger puzzle: Was this decline in amphibian populations occurring only in California, or was it part of some larger pattern? By coincidence, the First World Congress of Herpetology was scheduled to take place later that year in Canterbury, England. So Wake seized the opportunity to discuss his disturbing observations with other herpetologists. What he discovered, to his dismay, was that many of the attendees had witnessed the same phenomenon in scattered areas around the globe.

Wake took their reports and his own to the next meeting of the National Academy of Sciences Board of Biology, to which he belonged, and convinced its members to assemble a group of leading international amphibian experts to evaluate the evidence. The group, which convened in February 1990 in Irvine, California, quickly concluded that although most of the evidence for amphibian declines was anecdotal, the sheer number of widely dispersed informal reports indicated that the situation could be an environmental emergency, and that an international working group should conduct a full scientific investigation.

By the end of the year, after approaching several potential sponsors, Wake created the Declining Amphibian Population Task Force (DAPTF) under the aegis of the Species Survival Commission of the World Conservation Union, an international organization comprising more than 500 environmental groups including the U.S. Fish and Wildlife Service and the U.S. National Park Service. Based at the Open University in Milton Keynes, England, the task force recruited more than 1,200 scientists to determine whether declining amphibian populations will simply rebound as part of some normal cycle or whether they truly are disappearing from the face of the earth.

Why We Care About the Victims

One reason so many amphibian biologists were eager to join the task force was simply because they were worried they might be losing their object of study. But they were even more concerned for other reasons that everyone can appreciate. The first is the ethical consideration that amphibians have the right to exist. If people are responsible for amphibian disappearances, then people have a moral obligation to prevent them. Most religious traditions assign value to all living organisms. Even Judeo-Christianity, which espouses that humans are a special creation of God and are given dominion over the rest of the living organisms on earth, teaches that this relationship should be a stewardship, not a slaughter.

Second, amphibians are fascinating organisms that act in complex ways with each other and their environments. Consider the life history of the Central American strawberry poison frog, *Dendrobates pumilio*. At the beginning of their reproductive cycle, males call for females from perches on the tropical forest floor. After mating, the female lays her eggs in the forest's leaf litter. The father then revisits the eggs and keeps them moist with bladder water. When the eggs hatch into tadpoles, the mother carries them on her back and deposits each one into a tiny pool of water, often dew that collects at the base of bromeliad leaves. Because there is seldom enough food for even a single tadpole in these pools, the mother revisits each one every few days and lays an unfertilized egg for her offspring to eat. As the frogs mature, they synthesize poison toxins in their brightly colored skin from compounds found in the native arthropods on which they feed. If such frog species disappear, we lose valuable information about life on earth.

Third, amphibians may provide direct benefit to humans. One example is the gastric brooding frog, *Rheobatrachus silus*, of Queensland, Australia. After the female's eggs are fertilized, she swallows them and uses her stomach as a brood pouch, somehow switching off her digestive enzymes during the incubation period. Knowledge of such an enzyme-suppression mechanism might have proven helpful to people suffering from gastric ulcers. Unfortunately, while these and other biological aspects of *R. silus* were being investigated, the species disappeared from its natural environment and all specimens in the laboratory died. For a rough idea of what we'd be missing if many such species disappeared, consider some benefits that have already been realized, including a pain killer recently derived from poison-frog toxins and a nonirritating vaginal cream made from frog skin that prevents pregnancy and protects against sexually transmitted diseases.

The fourth and primary reason that the task force was established is that amphibians are important indicators of general environmental health. Because most amphibians have a biphasic life cycle—they spend their early stages in water and their adult life on land—and have extremely thin, permeable skin, any changes in either aquatic or terrestrial environments may significantly affect these creatures. Thus, amphibians may provide early warnings of deteriorating environments that appear unaltered to human perception.

Gathering Evidence

A concerted effort by the enlisted scientists has provided us with far greater documentation of amphibian decline than we had in 1990 when the task force was formed. One suspicion that researchers confirmed is that most amphibian declines and disappearances are directly related to habitat modification. Furthermore, when the habitat change is dramatic, so are the effects. For example, in the United Kingdom, where many—in some areas 80 percent [of]—breeding ponds have been filled in over the past 50 years, all six native amphibian species have suffered dramatic population declines. Elsewhere, along a well-studied area on Volcan Tajumulco, the highest mountain in Guatemala, only 1 of 8 species of salamanders was able to survive after cattle ranchers converted the upper cloud forest zone into grazing pastures. Herpetologists also discovered that seemingly modest changes in habitats can also have profound effects. For instance, to the casual observer, it would appear that the arroyo toad *(Bufo microscaphus californicus)*, whose habitat now exists entirely within uninhabited parks in California, is well protected. But the major streams that fed the best breeding sites have been dammed, and what remains of the stream bed plains is now being overrun by all-terrain sport vehicles. Because the larvae cannot live in the silty conditions that result from these modifications, toad populations have decreased alarmingly.

Perhaps the most disturbing finding, however, is that amphibian declines are occurring in diverse locations in relatively undisturbed habitats. Consider the following cases:

In Australia, herpetologists have known since the late 1970s that populations of *R. silus,* the gastric brooding frog, were declining in pristine sites. After learning at the First World Congress of Herpetology that the decline might be symptomatic of a worldwide problem, the Australians launched a campaign to inventory all known amphibian localities throughout their rainforests, and to initiate long-term monitoring programs in some key areas. The researchers had since counted 14 frog species from remote habitats whose once-abundant populations had either completely vanished or had been reduced to only a few frogs.

In California, biologists Charles Drost and Gary Fellers, both of whom are now with the U.S. Geological Survey, devised a clever approach to evaluate the status of amphibian populations in Yosemite National Park. Using extensive field notes of biologists Joseph Grinnell and Tracy Storer—who recorded detailed descriptions of the area's amphibian breeding sites between 1915 and 1919—Drost and Fellers were able to reassess the amphibian populations at the same sites. The fact that the researchers were able to relocate every site proved that no obvious change had occurred in the habitat during the intervening 75 years. Sadly, they also found that most of the amphibians were gone: whereas Grinnell and Storer counted 7 different amphibian species at 70 locations, Drost and Fellers could now find only 4 at 26 sites.

The elfin forests on the ridge crest at Monteverde, Costa Rica, have witnessed perhaps the most notorious disappearance of an amphibian population from an undisturbed habitat—that of *Bufo periglenes,* the golden toad. Among the world's most colorful amphibians, the brilliant golden males differ dramatically from the equally flamboyant black, red, and yellow females. Largely because of their spectacular beauty, golden toads—known to science only since the 1960s (although the Quakers who colonized the Monteverde area were aware of their existence before then)—served as the focus of concerted efforts to conserve the local habitat. In fact, a golden toad is depicted on the same sign with a panda to mark the entrance to a 328-hectare preserve established in 1972 by the Tropical Science Center of Costa Rica and the World [Wide] Fund for Nature. Later endeavors by other conservation groups tripled the size of the preserve to 10,500 hectares and finally more than doubled it again by adjoining it to the 16,000-hectare Children's International Rainforest.

Despite these conservation efforts, the golden-toad population crashed in 1988. During April and May of 1987, "more than 1,500 toads gathered to mate in temporary pools at Brillante, the principal known breeding site," report biologists Martha Crump and Alan Pounds, in the March 1994 issue of *Conservation Biology.* "But in 1988 and again in 1989, only a single toad appeared at Brillante, and a few others gathered 4 to 5 kilometers [to the southeast]. During 1990 to 1992," the researchers note, "despite our intense surveys, no golden toads were found." Nor have any been seen since.

In Puerto Rico, researchers have discovered that two species, including *Eleutherodactylus jasperi*—one of the world's few viviparous frog species (which, like mammals, produce live young instead of eggs)—have apparently become extinct though their habitat still appears suitable.

In Ecuador and Venezuela, eight species have been reported absent from the cloud forests of the Andes mountains. One genus in particular, the *Atelopus,* was once incredibly abundant (researchers could collect hundreds in an hour). But in 1990, Enrique LaMarca, a biologist at the University of the Andes in Venezuela—having spent more than 300 hours during 34 separate field trips searching for the frogs—reported finding only one specimen of *A. mucabajiensis* and two *A. soriani.* Another species in the genus, *A. oxyxrhynchus,* which LaMarca reported observing walking by the dozens on the forest floor, has not been seen since 1978.

In the Atlantic Forests of southeast Brazil, specifically at a well-studied site in Boraceia, São Paulo, seven common amphibian species disappeared in 1979. The site has since been revisited numerous times by several herpetologists including Jaime Bertolucci, a doctoral student at the University of São Paulo, who conducted an intensive year-long study of the ecology of tadpoles. But none of the species that disappeared in 1979 have ever been found.

Similarly well-documented studies have found amphibian disappearances or declines from relatively undisturbed habitats elsewhere in these and other regions, including the U.S. Rocky Mountains and the Cascade Mountain Range in Washington, Oregon, and California.

Possible Suspects

Though more work must be done to plug the gaps in our knowledge of amphibian declines, these studies allow us to draw an important conclusion: amphibian populations, in far-flung locations, are indeed disappearing even in seemingly virgin environments. The challenge, therefore, is no longer merely to preserve habitat, though that is still a vital task. We must also discover and address the less obvious reasons for the demise of these creatures as well as determine what fate they might portend for other species, including ourselves.

Prominent among the suspects thought responsible for declining amphibian populations, at least in specific locales, [are] agricultural chemicals and pesticides. In many parts of the world, certain amphibian species have thrived in agricultural areas, taking advantage of artificial water bodies used for irrigation and watering livestock. But the chemicals found in farmland breeding sites interfere with normal amphibian development. Michael Tyler, a biologist at the University of Adelaide in Australia and a board member of the Declining Amphibian Populations Task Force, explains that the problem with some herbicides is not the active ingredient itself, for example glyphosate, but rather a detergent additive that acts as a dispersant or wetting agent. The detergent breaks down the surface tension at the leaf surface to enable spray droplets to completely cover the leaf. However, the agent also interferes with respiration in frogs through the skin and even more so with respiration of tadpoles through gills. Michael Lannoo, a biologist at Ball State University, also points out that some pesticides such as methoprene (used for mosquito control) break down into a compound resembling [retinoic] acid, which has been shown in the laboratory to produce severe amphibian limb deformities that would render individuals incapable of escaping predators.

Other pollutants under investigation are being blamed for more regional amphibian declines. Among the leading culprits for these losses may be acid rain. In fact, researchers have found that almost all amphibian eggs or larvae tested so far cannot survive in water with a pH of less than 4.5. Yet acid rains, commonly in the 3.5 range, can lower the pH of ponds and streams from a normal average of about 7.0 to lethal levels. In fact, acid rain has been identified as a cause of amphibian declines in lakes and ponds in Canada, Scandinavia, and Eastern Europe.

Chief among the candidates likely to be responsible for amphibian declines on an even wider, perhaps global, basis is ozone depletion. Recent studies in Oregon have shown that rising levels of ultraviolet-B (UV-B) radiation resulting from the depletion of the earth's ozone layer have undermined the hatching success of eggs in some native amphibian species. The researchers suggest that other amphibians most likely to be affected by increased UV-B radiation—which, at elevated levels, breaks down the DNA molecule—are those living at cooler higher elevations and extreme latitudes, where the ozone layer is thinnest but where amphibians must bask in the sunlight to regulate body temperature.

Environmental estrogens may also be responsible for global declines. Researchers believe that these pollutants, which result from the chemical breakdown of pesticides such as DDT, are likely to severely affect the reproductive biology of amphibians, as they have been shown to do in other aquatic organisms, such as fish and alligators. In fact, in laboratory studies, Tyrone Hayes, an endocrinologist at the University of California, Berkeley, found that such environmental estrogens masculinized female Japanese tree frogs, *Buergeria buergeri,* and feminized male pine woods tree frogs, *Hyla femoralis,* causing both populations to become sterile. These estrogens, whose molecules do not break down easily in the environment, stockpile in silt on the bottoms of ponds and lakes, where they are ingested by bottom-feeding amphibian larvae. Some of these agents are effective in very small concentrations and are easily wind-borne, making them a global threat regardless of their point of origin.

Inconclusive Evidence

We must conduct more research to determine which, if any, of these factors are responsible for declining amphibian populations in relatively pristine habitats. One approach would be to compare undisturbed sites where amphibian populations are healthy to similar habitats where the populations are in serious decline. One such grouping exists in the Andes mountains in Ecuador, Colombia, and Venezuela. While amphibians continue to thrive in high-elevation habitats in Colombia, they have disappeared from virtually identical habitats in Ecuador and Venezuela. Might something as straightforward as introducing predators such as trout into the waters of Ecuador and Venezuela, but not Colombia, be responsible? Or might atmospheric transport of agricultural chemicals applied in lowland regions of Ecuador and Venezuela be causing problems? An elegant set of comparative studies and experiments could be designed to address such questions at these and other promising groups of undisturbed sites in lowland and cloud forest habitats of Africa, South America, southeast Asia, and Madagascar.

Another approach would include studies aimed at *rejecting* regional or global factors as causes of amphibian declines. Most research has tried to verify the link between reduced frog populations and factors such as high UV-B concentrations. But some studies suggest that UV-B, as a single factor, is not responsible for amphibian declines, since several species, such as the golden toad of Costa Rica, are never exposed to the sun's ultraviolet rays. In fact, golden toads lived underground all year long, except for a few days at the end of the dry season when they emerged to breed. But even then they were protected under the canopy of Monteverde's elfin forest, which (even though short by tropical lowland standards) effectively filters out the ultraviolet radiation. Moreover, because females chose to lay their eggs in well-shaded pools, the now-extinct golden toads were never exposed to UV-B even as eggs or larvae.

Such an analysis doesn't mean that rising UV-B levels are not killing off amphibians elsewhere. In fact, studies of amphibians exposed to such radiation are under way in the mountains of Chile and Argentina. It does, however, suggest that no single factor may be responsible for all declines. Perhaps more significant, the analysis also raises the possibility that more than one factor may be at play at each location. For example, if an amphibian population is subject to sublethal stresses from habitat fragmentation and acid rain, might it be more likely to succumb to an additional stress from some regional or global factor such as climate change or estrogen mimics?

Some research shows that such scenarios are possible. A study of the western toad *Bufo boreas,* common to the Elk and West Elk Mountains of Colorado, serves as one example. Cynthia Carey, a biologist at the University of Colorado, who began studying these toads in 1974, discovered that they had contracted "red leg" disease, a normally nonfatal illness caused by *Aeromonas hydrophila,* a naturally occurring bacterium. Over the next eight years, Carey found that the toads, once common in the mountains, had almost completely disappeared. Her conclusion was that some environmental factor, or the synergistic effects of several factors, may have caused the toads to secrete elevated levels of hormones that compromised their immune system and led to their infection and eventual death.

Studies such as these demonstrate that the underlying causes of amphibian declines may be far more complex than anyone originally imagined. Thus, studies that examine possible synergistic effects and help us tease out the relative contribution of each must be among our research priorities.

Interim Recommendations

Though much research lies ahead, we can take some practical steps immediately to halt the decline of amphibian populations. Perhaps the most obvious is to preserve remaining amphibian habitats. One novel approach would be to consider the health of amphibians in environmental impact assessments. In fact, this practice proved highly successful at a highway-construction site in British Columbia recently. Typically, whenever highways are built in the forested Canadian province, workers create roadside ditches and scour them of all vegetation. But in this case, thanks to a herpetologist included on the environmental-impact study team, the road builders added parts of fallen trees to the ditches, enabling native amphibians to use them as breeding sites.

Another simple but valuable step would be to consider amphibians in environmental assessment programs as bioindicators of overall ecosystem health. Because the eggs of many amphibians lack a protective covering and are laid at or near the surface of a body of water, they are very sensitive to both air- and water-borne pollutants. Also, because the climatic factors typically determine the onset, duration, and intensity of amphibian mating activity, careful monitoring of breeding populations can provide an extremely sensitive assay of climate change.

Finally, the latest findings regarding causes of amphibian declines need to be communicated both to international policymakers, who are in a position to set research priorities and fund additional studies, and to the public at large, which can influence their decisions. Americans are now much more aware of issues concerning amphibians than they were even a decade ago, thanks in large part to a number of excellent television documentaries that have focused on dwindling amphibian populations. But scientists and the media must continue to spread the word to convince people around the world that these precious creatures are worth their concern.

Stephen Leahy

"The Sound of Silence" (1998)

It was a muggy, buggy late-June evening when I went north hoping to find one of nature's spectacular performances—a choir of bullfrogs *(Rana catesbeiana)* singing their mating songs among the fireflies. During warm June and July nights, males call with deep, resonant "brumm, brumm, brumm" sounds to attract females. And the melody is loud enough to be heard a kilometre away.

"To be in the middle of a chorus of males is just an extraordinary experience," says Michael Berrill, a biologist studying bullfrogs at Ontario's Trent University. But 75 kilometres north of Toronto on shallow, weedy Lake Scugog and its several hundred acres of wetland, we heard only the whine of pancake-flat bass boats, the hissing of car tires, and the general bustle from the small settlements around the lake. While mosquitoes buzzed loudly and fireflies sparked in the dark, the plump, green "canaries" weren't there to sing any more.

Bullfrogs haven't been singing as they once did in eastern Ontario either. Reports of their disappearance by residents and cottagers have been supported by the very few population surveys done. These have shown recent declines of 60 percent in Lanark County and 90 percent in one Peterborough County study area. After being advised that people were "harvesting" the noisy and easy-to-catch breeding males, Ontario's Ministry of Natural Resources (OMNR) banned the collection of bullfrogs for commercial purposes in 1995, and personal use in 1996.

No one knows if the bans have been successful. "We don't have any idea what the level of illegal harvest is," admits OMNR biologist Shaun Thompson, whose downsized ministry has no money to study bullfrogs or staff to be "frog police." However, he points out that removing even a few dozen adults could silence a small lake; and the loss of large numbers could have a devastating effect on entire populations. (The one and only bullfrog poacher who's been caught and charged had 2,000 adults in his possession. They were released unharmed.)

No one harvests bullfrogs from Lake Sasajewun—a 63-hectare, cool, blue-green lake set amid the rocky, pine-clad ridges of Ontario's Algonquin Park. No one except Ron Brooks, professor of zoology at the University of Guelph, and his students. Each summer, at the park's field research station, they collect frogs and turtles to observe before returning them to their natural habitat.

"Bullfrogs are at the northern end of their range here," notes Brooks, who has been studying the area's bullfrog populations for the past 12 years. While the amphibians tend to be smaller and fewer in number than in the past, the white-haired professor says they seem to be doing fine.

After months behind the lectern, field research is pretty tiring even for a vigorous and seasoned researcher. Especially when species like bullfrogs are active mostly at night. But this afternoon, at the height of the breeding season, a male bullfrog sits in the warm, shallow water at our feet. Brooks explains that, because "chorusing or singing all night uses a lot of energy," the frogs need to feed as much as they can during the day.

The bullfrog's appetite is legendary in terms of variety and size. While insects are the mainstay, the predatory amphibian will go after practically anything it can cram into its maw, swallowing prey up to three-quarters its own size. Snakes (rattlesnakes included), mice, fish, turtles, worms, crayfish, tadpoles, and frogs (even smaller bullfrogs) are all on the menu. Small birds and bats are fair game as well.

Like all frogs, *Rana catesbeiana*'s hunting tool is a relatively short tongue attached to the front of its mouth. Equipped with mucous glands, the appendage exudes sticky secretions to catch most prey. Bullfrogs patiently wait until their dinner is close; then, in a literal blink of an eye, they leap toward it, flipping their tongue out and up, curling the tip around their victim. The tongue keeps the bullfrog's prey stuck long enough to be folded back into the mouth where it is held by pressing it against teeth on the roof of the mouth.

Then the tongue and other muscles push it toward the oesophagus. The bullfrog's eyes sink downward during the swallowing to help force the food into the digestive tract.

But the hunter is also hunted, notes Brooks. Raccoons, snapping turtles, and otters enjoy adult bullfrogs, while juveniles and young froglets have to avoid herons, mink, bass, pickerel, and pike, among others. Last winter, according to the professor, otters feasted on bullfrogs overwintering on the bottom of Lake Sasa. He estimates "we lost 80 percent of the males, mostly because of otters."

Although bullfrogs are powerful enough swimmers to evade these predators, the very cold water under the lake ice slows them to a crawl. Amphibians are ectothermic—i.e., they depend on external sources of heat for their body temperature—and their metabolism slows when the temperature drops. In winter, it slows so much they can survive for months on a lake bottom buried in the mud or leaf litter without food or need to surface for air. Not possessing gills, they get the little oxygen they need from water through their thin, blood-vessel-rich skin.

This kind of skin-breathing also happens inside a frog's mouth, which is why they often have their mouth open under water. With rising temperatures in spring, their reviving metabolism needs more oxygen and their lungs kick in. The throat acts as a kind of bellows, pumping air in and out through the nostrils and into the mouth where the oxygen is absorbed through the skin. Once or twice a minute, the bellows action stops momentarily and there's a slight twitch around the waist as the frog takes air into its lungs.

Ron Brooks and I watch the male at our feet breathe while deer flies buzz happily around us in the warm afternoon sunshine. He tells me that his graduate student, Kevin Judge, estimates there are over 100 males calling at night on the lake. To determine the number, he caught, tattooed, and microchipped them all, says the professor. He adds, "You can ask him [about his research project] when he wakes up."

We find a sleepy Judge having breakfast by himself in the cookhouse at four in the afternoon. With help from undergrad Sarah Swanson, he spends six to eight hours every night paddling (as well as dragging a canoe over logs and through muck because of this year's low water levels) in order to catch the singing males. "We'll probably catch 40 to 50 tonight," he says with a stifled yawn.

Judge is doing his master's thesis on male bullfrog behaviour. He hopes to learn if certain males are more successful breeders than others and, if so, why. The most likely breeders are the biggest males at six or seven years of age. They try to attract females by calling them to their one-to-two-metre circular territory, explains the graduate student.

These singing males are usually found in a group, in what's appropriately called a chorus. There can be a number of competing chorus groups depending on the size of the male population. Each chorus competes to draw females to their group, and like many an amateur choir, each singer attempts to outdo the other.

Sometimes things get physical. With a loud "hick," a pair of aggressive singers engage in furious wrestling matches, often looking like a pair of straining sumo wrestlers, nose to nose, arms flailing. Eventually one topples the other, sometimes holding the loser underwater for a time and then chasing him away from the conquered territory.

Females cruise by these noisy groups and eventually sidle up to a male with a resonant voice and well-chosen territory. When ready, she gives him a nudge. With that encouragement, he climbs on her back [and] wraps his forelegs tightly around her in what is termed amplexis. She begins to deposit her eggs on the surface of the water while he fertilizes them. A mass of 6,000 or 7,000 eggs is usually laid in a thin sheet, but very large females can lay 20,000 eggs.

The couple soon parts, the female to rest, the male returning to his territory in hopes of another conquest. While size does matter to female bullfrogs, smaller and younger males sometimes slip unnoticed into a large dominant male's territory. They lie low in the water quietly waiting for a female to respond to the big male's call. Intercepting her, they sometimes carry off a successful impersonation, and pass their genes along to the next generation.

The newborn tadpoles are equipped with gills and spend their lives hiding from predators like game fish while scraping bacteria and microorganisms off plant stems. In Algonquin, as in other parts of Canada, bullfrogs remain in the tadpole stage for three years. It seems our short summers don't allow enough time for them to reach their full 12- to 14-centimetre size. (In contrast, adult bullfrogs in the southern U.S. are almost twice as large, and metamorphize the summer following their birth.)

Despite the huge numbers of eggs each female deposits, only one or two are likely to become adult bullfrogs. Leeches and small fish gobble the eggs up over the two-week incubation period; and they are affected by many environmental factors—including heat, cold, and acidity of the water—as well. There is also some evidence that the increase in ultraviolet light from the thinner ozone layer may have an impact on them.

Tadpoles, too, are very sensitive to environmental contaminants. Trent University's Michael Berrill says, "We've looked at a dozen or so pesticides and herbicides. . . . Bullfrogs, like other frogs, are very sensitive to pesticide contaminants, at disturbingly low levels." Young tadpoles are the most vulnerable, suffering paralysis and "a high mortality at quite low levels of exposure."

Even those that survive don't feed properly and their growth is stunted, Berrill found. The chemicals he tested are those most commonly used in agriculture, horticulture, forestry, and by household gardeners. "Even [with] the herbicides that are supposed to be relatively safe, we still get massive problems," he notes.

Researchers in various locations have discovered frogs with extra eyes, split limbs, missing digits, and other deformities. The first wave, which occurred in the late 1970s, was linked to acid rain. The most recent outbreak was first noticed in 1995 by some Minnesota school children. Now investigators have found that much of that state as well as Wisconsin, Quebec, and Ontario have ponds in agricultural areas where 65 to 95 percent of the frogs are deformed.

Les Lowcock, research associate at the Canadian Centre for Biodiversity and Conservation Biology, is convinced that these large-scale mutations are a result of exposure to pesticides during the developmental stages of the frogs' life cycle. His recently published research reveals that frogs from pesticide-contaminated ponds in Quebec cornfields have mangled DNA. "Their chromosomes are heavily damaged, [with] pieces missing and parts shuffled around," he says.

More chilling is the fact that even the remaining healthy looking frogs in these ponds carry damaged DNA. "We don't know what impact this will have down the line," admits Lowcock. While he suspects this latent genetic damage will result in hormonal or neural effects on succeeding generations, he notes "there is no research on the long-term effects."

Lowcock is concerned that what's happening to frogs also might be happening to people. Some biological systems—such as the endocrine system—are virtually identical between species, including humans and amphibians. The endocrine system employs hormones as messengers to control development of the brain, sex organs, and immune system. Many pesticides and herbicides, including the commonly used atrazine, can disrupt this messaging system, resulting in reduced fertility, impaired immune systems, and compromised neural function.

However, Lowcock doesn't attribute declines in bullfrog or other amphibian populations to environmental contaminants alone. Like Berrill, Brooks, Judge, and others, he believes there are a number of factors involved. As David Green, the current co-chair of the reptiles and amphibians section for the Committee on the Status of Endangered Wildlife in Canada (COSEWIC), says, "It's very likely lots of little things all adding up."

While commercial and personal harvesting for the kitchen pot once headed the list of threats to bullfrogs, loss of habitat through shoreline development and draining of marshes and ponds has probably taken its place. All frogs need aquatic and shoreline vegetation—cement breakwaters and lawns are biological wastelands for amphibians and many other species. Roads are even less hospitable. Bullfrogs often travel short distances overland in search of better habitat. On warm rainy nights, frogs of all kinds are squashed by the thousands on roadways.

Then there are the small froglets and juveniles that are used in large numbers as bait and by the pet trade. Not to mention the tadpoles that disappear when lakes are stocked with hungry game fish like trout, muskie, and pike.

Fluctuating water levels from dams can also leave eggs high and dry, while declines of bullfrog food sources can make the amphibians vulnerable to disease or unable to reproduce.

As well, frogs are susceptible to fungal infections, especially in a weakened condition. Toss some pesticides, gasoline, pulp mill waste, or other toxins into the water, and that could mean the end of bullfrogs in that pond or lake. Given all of the above, it's not surprising that much of southern Ontario is silent today.

Although a lot more research is now being done on various frog species, the prognosis is not good: "The decline in amphibians will continue and many species will go extinct," predicts Green. However, the bullfrog will probably avoid the fate of the northern leopard frog and the Great Basin spadefoot toad, which are recent additions to the COSEWIC list. Years ago, bullfrogs were introduced into western North America—primarily for the frog-leg trade— and they've done well, particularly in British Columbia. So well, in fact, that they're eating their way through native frogs and other species, putting them in danger of extinction.

California and Colorado researchers looking into the decline of native frog populations are also fingering the introduced *Rana catesbeiana*. Aside from its frog-gobbling appetite, the main reason for the bullfrog's success in the west is that its foul-tasting eggs and tadpoles are only palatable to a few fish species.

So why does it matter that the bullfrog isn't where it used to be, as long as there are lots around somewhere else? For one thing, its absence in a given area has an impact on the local food web. For another, the decline of this plump green canary could signal trouble ahead for the humans who share its environment. And finally, of considerable importance (at least to me) is the lost opportunity of hearing that extraordinary basso profundo chorus.

VIRGINIA MORELL

"Are Pathogens Felling Frogs?" (1999)

Townsville, Queensland, Australia—On a Tuesday morning last August, Ken Aplin, curator of reptiles and amphibians at the Western Australia Museum in Perth, got a chilling phone call. The owner of a nearby organic nursery want-

---

Reprinted with permission from *Science* 284 (30 April 1999), 728–31. Copyright 1999 American Association for the Advancement of Science.

ed to know why hundreds of frogs—common motorbike frogs, whose call sounds like a motorcycle changing gears—had suddenly died on his chemical- and pesticide-free grounds. Aplin didn't have a ready answer, but he feared the worst. A few weeks later, his suspicions were confirmed: A colleague tested a freshly dead motorbike frog from another nearby source and found that the animal was lethally infected with a parasitic chytrid fungus—a virulent amphibian pathogen that had caused sudden, massive die-offs of more than a dozen frog species here in Queensland; four species have apparently gone extinct. Now, it seemed, the fungus had leaped 6000 kilometers and across the dry Nullarbor Desert, spreading to western Australia, a region with a rich endemic frog fauna that had never seen massive die-offs before.

"It was like hearing about a first case of cholera," says Aplin. "I feared it would spread." In the next few months, the disease was found on other dead frogs in Perth and nearby towns to the south, often killing every frog in backyard ponds. "I suspect that we're on the edge of a major outbreak that will cause mass mortalities in the next few months, when frogs gather to breed," he adds.

Australia considers itself the front line in dealing with this frog pathogen, *Batrachochytrium dendrobatidis,* a new genus and species described as a lethal disease 9 months ago (*Science,* 3 July 1998, p. 23) and named only last month. But the chytrid is not just an Australian problem. It is suspected in the catastrophic disappearance of frogs in Panama and Costa Rica; and it is implicated in mass die-offs in the United States as well. Indeed, some researchers—mostly epidemiologists—say that this virulent new fungus may be the key factor in the sudden, mysterious decline of frogs around the globe, particularly those from wilderness areas of the Americas and Australia.

Since the 1970s, populations and species of frogs have been vanishing worldwide, and deformities such as missing legs have been turning up with alarming frequency, sparking massive research and monitoring programs. In many cases, frog populations or even entire species in pristine, remote mountain areas suddenly vanished in a few months. Baffled about why frogs in protected areas would be so vulnerable, many researchers have looked to the global environment, arguing that frogs are like canaries in a coal mine, serving as indicators of global ecological health. They have studied a plethora of environmental suspects, ranging from increased ultraviolet (UV) light to global warming and wind-borne pollutants. Yet despite almost a decade of intense research and some loose correlations between die-offs and environmental factors, no one so far has been able to show that these factors are actually killing frogs.

Thus, when the chytrid was first fingered, many scientists regarded it as just one of many "smoking guns" that would prove less convincing as time went by. But this time, researchers have bodies to prove the case. The Australian experience has galvanized researchers there, and now scientists elsewhere are taking seriously the idea that the chytrid plays a central role in the declines. A team of U.S. researchers has just proven in the lab that the chytrid alone can kill healthy frogs. And by studying preserved specimens, other researchers have now implicated the fungus in some of the very die-offs that first raised the amphibian alarm in the United States, including mass deaths of leopard frogs, *Rana pipiens,* in the Colorado Rockies back in 1974, and in more recent disappearances of Arizona's lowland leopard frogs. "It's increasingly clear that we need to treat the chytrid—and amphibian diseases in general—as a serious threat," says Cynthia Carey, a physiological ecologist at the University of Colorado, Boulder. "Diseases are killing frogs and we need to know why."

Still, when it comes to worldwide frog declines, several leading U.S. herpetologists, such as David Wake of the University of California, Berkeley, resist the notion of a single cause; others note that some chytrid-infected frogs survive. So researchers are working to test two competing ideas: that the chytrid is an emerging pathogen sweeping through previously unexposed populations, or that an environmental cofactor such as increased UV light or climate change is magnifying the chytrid's effects. "I don't think you can rule out" such cofactors, says Donald Nichols, a pathologist at the National Zoo in Washington, D.C., who first identified the disease. He and others hope that a flurry of chytrid research will help them find out.

Death Down Under

The story down under starts here in northeastern Australia, where herpetologists began monitoring frog populations in 1989 after several species dwindled or disappeared from some of the least touched places on Earth, including World Heritage rainforest parks in the Atherton Tablelands. At first, environmental pollutants were thought to be the cause. But the streams where the frogs died aren't polluted, and the amount of UV light here hasn't risen during the past few decades. In this case, "you can rule out any of these environmental cofactors," says Richard Speare, an infectious disease specialist at James Cooke University here. And infected frogs don't show "multiple opportunistic infections, such as you'd expect if their immune systems were compromised," notes parasitologist Peter Daszak of the University of Georgia, Athens, who identified the chytrid in Australian frogs.

The pattern of death "has all the hallmarks of an emerging pathogen," says Speare, particularly its ability to infect a broad range of animals; thus it can continue to spread even after wiping out one species entirely. In Australia, researchers have found the chytrid on almost every suffering frog population they have checked, and they have been able to chart the fungus's spread through the continent, starting at the northern end of the epidemic in Queensland in 1993. So far they've traced it 300 kilometers south and 4 years back to 1989. The very first die-offs struck just north of the port of Brisbane in 1979, so Speare speculates that the fungus was introduced to Australia in the late 1970s, perhaps on an exotic frog.

Traveling at about 100 kilometers a year, the chytrid decimated many species, moving "like a wave" through frog populations in successive localities. Last year it apparently crossed the continent to Perth and western Australia, perhaps on the feet of an infected frog that stowed away in a box of fruit destined for Perth. "That happens all the time," says Aplin. "A store clerk opens a box of bananas from Queensland and out hops a frog"—and with it the chytrid. Such a scenario would explain why the first cases in southwestern Australia appeared in urban areas.

Not all Aussie frogs have suffered equally from the chytrid: Queensland's green-eyed tree frog, which died from the disease in great numbers in the late 1980s and early '90s, is slowly coming back, suggesting that this species may have some resistance. But four stream-dwelling species, including two unique gastric-brooding frogs, are presumed to have been wiped out by the fungus. It has also infected another 24 species and drastically shrunk the population of 11 of these.

At the same time the Australians were identifying the pathogen, researchers in Central America were puzzling over their own frog deaths. After hearing about the Australian situation, researchers checked for the chytrid—and found it on 10 different dead frog species in western Panama. The chytrid is also linked to the disappearance of numerous species in the rainforests of Costa Rica, says herpetologist Karen Lipps of Southern Illinois University in Carbondale.

Preliminary genetic work suggests that the Central American frog fungus is the same one that plagued Aussie frogs. David Porter from the University of Georgia, Athens, chytrid specialist Joyce Longcore from the University of Maine, Orono, and graduate student Timothy James from Duke University in Durham, North Carolina, have found that the 18S ribosomal DNA genetic regions of chytrids are nearly identical, implying that a single fungal species is sweeping into new realms worldwide. James also notes that it is "quite an odd fungus," different from the chytrids found commonly in soil and water. "There's no doubt we're seeing a new, emerging disease, one that is highly pathogenic and hits a wide range of amphibians," concludes Daszak. Further molecular data should reveal how recently the chytrid has spread through frogs on various continents, and perhaps whether it has newly evolved to attack amphibians or if it is an old frog nemesis now invading naïve populations.

Some researchers haven't been sure that the chytrid can kill healthy frogs. But a few months ago, in as-yet-unpublished work, a team led by the National Zoo's Nichols isolated the chytrid from poison dart frogs killed by the disease. They then inoculated healthy frogs with this chytrid. All infected frogs died, whereas frogs inoculated with a placebo did not. Next, they reisolated the chytrid from the second batch of dead frogs, a sequence of experiments that fulfills what are called Koch's postulates, the gold standard for proving that an organism causes a disease.

The chytrid apparently uses the keratin in the frog's skin as a nutrient. Its motile, water-borne spores invade surface skin cells and grow and divide there asexually. No one is quite sure just how it kills frogs; Speare suspects that the fungus secretes a toxin, as dying frogs are unable to keep their balance and seem to have seizures, whereas others think the fungus blocks water uptake. In Australia, species hit hardest tend to spend most of their lives in water and live at higher and cooler altitudes, says Speare. That fits with both Australian and American lab studies showing that the chytrid is hard to grow at above 30°C and dies without water.

Revisiting U.S. Die-Offs

The chytrid's cold, wet-loving habits may help explain dramatic frog deaths in the United States as well, some researchers say. Take the decline of the once-common lowland leopard frog in Arizona. Over the last decade, researchers have pointed their fingers at the usual list of suspects, including loss of wetlands, heavy metals, and bacterial infections. But no one had ever seen wild frogs dying en masse from these killers. Scientists only knew that when they returned to the field each season, fewer and fewer frog populations were left.

Then, in January of this year, Michael Sredl, a herpetologist at the Arizona Game and Fish Department in Phoenix, spotted a leopard frog population north of Phoenix in its death throes: The frogs were emaciated, trembling, and rigid. They had no obvious skin lesions, ulcers, or fungal growth, but histological sections of their skin revealed lethal numbers of the chytrid and its spores.

This sighting also taught Sredl something else: Frogs were dying in winter. "We'd missed it every other year, because herpetologists usually don't go looking for their animals in the winter," he says. The chytrid has now been "positively implicated" in die-offs of two leopard frog species in Arizona, as well as a species of tree frog, says Sredl. It's "under investigation in the declines of all Arizona ranid frogs." And it's been found in specimens collected last year of Pacific tree frogs and endangered mountain yellow-legged frogs from California's High Sierras. Other scientists believe the chytrid may be responsible for a slew of other die-offs as well, including the extinction of *Rana pipiens* and *Bufo boreas* from the Colorado Rocky Mountains in the 1970s, a '70s crash of the *Rana pipiens* population around the Great Lakes, and sudden die-offs of ranids and toads from Wyoming to New Mexico in the 1980s.

Pathogens have been suggested as culprits before. Back in 1993, Norman Scott, a zoologist at the U.S. Geological Survey in San Simeon, California, suggested that a novel pathogen was killing western frogs. He named the disease the "postmetamorphic death syndrome," because newly metamorphosed frogs seemed to die overnight. That's a characteristic of chytrid infection, because tadpoles carry the disease only in their mouths and survive; after metamorphosis, the fungus spreads throughout the frog's skin and kills it. But Scott's idea received little attention at the time, perhaps, he says, because researchers were so determined to find an environmental cause.

Now U.S. scientists are gearing up to have both historical and freshly collected specimens checked for the fungus, something the Australians began a year ago. "Only a week ago, I would never have thought to do this—collect frogs for testing," says Michael Lannoo, a herpetologist at the Indiana University School of Medicine in Muncie, who recently realized that the pattern of declines he's studying in the Midwest's northern cricket frog matches that of earlier chytrid die-offs.

However, even the most avid defenders of the chytrid hypothesis say it's not responsible for every decline. The fungus has not been found in Europe, for example. And herpetologists generally agree that the biggest problem for frogs worldwide is habitat loss; a species with only a few small, fragmented populations is likely to be much more vulnerable to the chytrid or another disease. "When you turn a diverse ecosystem into a hog lot," notes Val Beasely, a veterinary toxicologist at the University of Illinois, Urbana-Champaign, who has found mild cases of chytrid fungus but no mass deaths in southern cricket frogs, "it's not a surprise that you get a disease."

What's more, many researchers, noting that there are correlations with environmental factors, still favor this kind of explanation. "It's way too early to rule out these other factors, such as pollutants and UV light," says Andrew Blaustein, an ecologist at Oregon State University in Corvallis. Veterinary researchers such as Nichols add that most fungal infections in amphibians are opportunistic, moving in when animals are already stressed or injured from other sources. And in the United States, there's no clear pattern yet of deaths spreading out from one locality, as in Australia.

Wake also questions whether one fungus could kill so many different kinds of amphibians—frogs and toads in 19 families have died from it, and it has infected salamanders too. "I don't know of any other pathogen that kills like that; for example, one that kills all mammals," he says. In his view the genetics are too preliminary to be sure that the chytrid is the same species and strain in Australia and the Americas. Until that has been shown, he says, "any suggestion of an epidemic [is] irresponsible." Even Nichols, who proved that the chytrid can fatally infect frogs, says to "count me among the skeptics who wonder what role the chytrids are playing" in the wild. He and others question whether something hasn't changed in the frogs' environment to weaken their resistance and promote the chytrid.

While U.S. researchers argue about the chytrid's role, in Australia, Speare, Aplin, and others are waging a campaign to try to stop it. They're tracking the fungus's spread, identifying susceptible species, and planning captive breeding programs. And even skeptical U.S. researchers are urging field precautions, in case herpetologists themselves are spreading the fungus via wet boots or collecting gear. Chytrid specialist Longcore, who named the new genus, notes that she brought a non-disease-causing type of chytrid from Puerto Rico home to Maine in the wet mud on her boots. Despite all such efforts, Speare and others fear that in Australia, the disease "will be spread like the plague" through new populations. At least, Aplin says, this time scientists will be able to watch one of these sudden declines in action, rather than discovering it after it's all over: "We've caught it this time close to the beginning." And that may provide answers to the many questions that still surround this strange frog killer.

JOCELYN KAISER

"A Trematode Parasite Causes Some Frog Deformities" (1999)

It was a disturbing sign that something might be going terribly wrong in the environment: Frogs with extra legs, missing limbs, and twisted jaws were popping up in ponds across the country. First spotted by schoolchildren in Minnesota in 1995, the famous malformed frogs, together with reports of declining frog populations worldwide, sparked concerns that the animals might be falling victim to some type of environmental degradation—a change that might even threaten human populations. The discoveries touched off a million-dollar-plus hunt to find the culprit, whether natural or humanmade. Two [recent studies] now point to a natural cause for at least some of the frog abnormalities.

A team led by a recent Stanford graduate [has obtained] results indicating that infection by a trematode, a kind of parasitic flatworm, is at fault. The researchers based this conclusion on experiments in which they showed that they could exactly duplicate the kinds of limb abnormalities and other deformities seen in California by infecting tadpoles with the trematode, which goes by the genus name *Ribeiroia*. Some experts say that this work, together with a second study that may exonerate certain chemicals suspected of causing the abnormalities, has now elevated parasites to the top of the list of possible causes of frog deformities across the country. "This is *the* best experimental evidence showing a cause for the limb deformities in amphibians," says Andrew

---

Reprinted with permission from *Science* 284 (30 April 1999), 731, 733. Copyright 1999 American Association for the Advancement of Science.

Blaustein, an ecologist at Oregon State University in Corvallis, who has studied whether ultraviolet light could explain the deformities.

Others caution, however, that *Ribeiroia* infections may not explain the different patterns of frog deformities seen outside of California, especially in the Midwest. "I do not believe that there's a single cause" for the deformities, says herpetologist Mike Lannoo of Indiana University School of Medicine in Muncie. Still, many experts are saying that after several years of frustration it's a relief to finally get some hard evidence for what might be happening to the frogs.

Since the first malformed frogs made headlines in Minnesota, deformities in at least 12 species of frogs and salamanders have been reported in Canada, Vermont, and 32 other states, often at rates of 8% or more, much higher than the rate of 1% or less expected in healthy populations. Investigators have pursued three main theories about what might be causing the problems: chemicals such as pesticides, increased ultraviolet light because of ozone destruction, or parasites (*Science,* 19 December 1997, p. 2051).

Despite the flurry of activity, however, no lab had grown a batch of frogs under environmentally relevant conditions and produced the same deformities seen in wild specimens of the same species—until now. Pieter Johnson began this project 2 years ago for his undergraduate thesis at Stanford, with ecologist Paul Ehrlich as his adviser. Johnson investigated some ponds about 45 minutes south of Palo Alto where up to 40% of emerging Pacific tree frogs had deformities, mostly extra, partial, or missing hindlegs. The water tested free of chemical pollutants, but he noticed that the ponds with deformed frogs always had planorbid snails, a first host for *Ribeiroia* trematodes. "That was a pretty substantial clue" that trematodes might explain the deformities, Johnson says.

That idea fit with a proposal developmental biologist Stanley Sessions of Hartwick College in Oneonta, New York, had made years earlier. In work published in 1990, Sessions had shown that he could induce extra legs in salamanders by implanting beads in their developing limbs, presumably because the beads move cells around. Noting that the cysts formed in infected hosts by trematodes could exert the same kind of mechanical forces as the beads, Sessions suggested that the worms could also cause limb deformities.

By the time he graduated last June, Johnson had dissected hundreds of frogs and found that they did in fact have trematode cysts clustered around their extra limbs. But he hadn't done any experiments exposing tadpoles to the parasites. "I couldn't let go that close" to a solution, he says. So he teamed up with two friends, Kevin Lunde and Euan Ritchie. They all spent the summer "working pretty intensely," Johnson recalls, often from 10 P.M. to dawn so they could catch the parasitic worms when they emerged from the snails and use them to infect the frogs.

After several false starts, the team began infecting tadpoles with *Ribeiroia* and watching them develop into adults. The results were "almost painfully textbook," Johnson says. Higher doses of the trematode produced more deformities, and the mix of multiple legs, partial and missing limbs, fused skin, and other oddities was very close to that seen in the frogs in the field. Johnson thinks the cysts may cause deformities by changing the positions of cells in a developing limb, as Sessions' beads apparently did, and may also produce some chemical that mimics a hormone.

The Johnson team's findings don't mean that some chemical in the environment couldn't be at work too, but in the accompanying report, Sessions offers evidence that seems to rule out at least one type of chemical that has been linked to the frog deformities: retinoids. Sessions compared the abnormalities in 391 preserved, multilegged Pacific tree frogs from California and Oregon to those known to be induced in the lab by retinoids. More than 90% of the time, for example, the chemical produces a "proximal-distal duplication," such as a new limb coming out of the elbow rather than the shoulder. The retinoids also cause only certain mirror-image limb duplications. Although Sessions found many specimens with other kinds of mirror-image duplications, none had proximal-distal duplications. "Retinoic acid gives you particular morphologies, and we just don't see that with the frogs," says Sessions.

Developmental biologist David Gardiner of the University of California, Irvine, who has been studying retinoids as a possible cause, disagrees, saying they are still in the running. "What the published literature says retinoids do and don't give you," he says, isn't clear-cut. Other researchers say that differences in the abnormalities seen in midwestern and eastern frogs also point to other causes besides parasites. Few have the extra legs seen in California, for

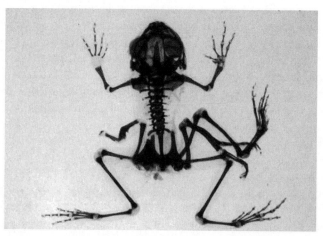

Figure 14
Pacific tree frog *(Hyla regilla)* with trematode-induced deformities
(Courtesy Stanley K. Sessions, Hartwick College)

example. And although some of the animals have cysts, so far nobody has found *Ribeiroia* in the midwestern frogs. In addition, Carol Meteyer, a wildlife pathologist at the U.S. Geological Survey in Madison, Wisconsin, says she has dissected hundreds of metamorphosing tadpoles from the affected ponds, and the cysts she has found do not appear until after the frogs' limb buds had developed—too late to do the damage Johnson describes.

But Lannoo and many others think parasites should be looked at more closely, even in those locales where chemicals are also suspected. "I don't for a minute think this is going to explain everything," says David Wake, director of the University of California, Berkeley's Museum of Vertebrate Zoology. But he adds that it's "a warning not to put all of your eggs in one basket" when trying to pin down the cause of the frog deformities.

GORDON L. MILLER

"Dimensions of Deformity"

The story by now has become almost legendary. In August 1995, a group of Minnesota middle school students discovered a passel of deformed frogs on a nature walk to a nearby field and pond. Recognizing that the percentage of frogs with missing legs, extra legs, or other malformations was apparently much higher than would occur in the normal course of natural events, they suspected that agricultural pesticides might be a factor. Environmental authorities were notified, the media got involved, Web sites were formed, and the world, or at least a significant portion of it, took notice. Since then, deformed amphibians have been found at many and varied sites throughout the United States and at some locations elsewhere. Researchers in both hip waders and lab coats are still busy teasing out evidence and testing hypotheses to explain this apparently variegated phenomenon.

When several scientists reported in the spring of 1999 that a tiny parasitic trematode was the immediate, if not always the ultimate, cause of at least some of the deformities, many members of the herpetological community breathed a collective sigh of relief. Finally, a piece of the puzzle had fallen rather solidly into place. But a whole range of researchers, including those who presented the findings, have emphasized that the puzzle is far from being fully solved. For example, trematodes can account for many cases of extra legs, but not nearly so well for missing legs or various other deformities that comprise the bulk of malformations around the country. For many cases of this type, scientists see chemical pollutants or perhaps increased ultraviolet radiation as more likely culprits. There is also the closely related question of a possible increase in the number of aquatic snails that host the trematodes before they move on to tadpoles. Such snails thrive in lakes with accelerated eutrophication caused by nitrate and phosphate pollution. The patterns in this puzzle are apparently so subtle and intertwined, and the pieces so widely dispersed, that there will be no easy resolution of them into a clear picture of the problem.[1]

It is interesting to notice, however, some of the responses evoked by the trematode reports. Within days, outspoken advocates of free enterprise seized upon the findings and began trumpeting trematodes as an embarrassment to environmentalists who had been suspecting anthropogenic causes for the deformities. One of these champions of commerce and industrial society has proposed that "the hysteria" caused by depicting frogs as poster children in a pollution campaign demonstrates "not the corruptions of capitalism, but the absurd lengths environmentalists will go to [to] manufacture a crisis in hopes of more government control." That writer also believes that what motivates environmentalists "is not a genuine concern for nature, but a hatred of man and his civilization."[2] Conveniently ignoring the facts that the parasitic habits of trematodes cannot explain all deformities, that trematodes may be cofactors with other environmental conditions, and that deformities are only a part of the much larger and still unsolved puzzle of amphibian decline, these evangelists of economic growth have seen this portion of amphibian research as a green light for business as usual.

The most interesting thing about this rush to exonerate humankind in the matter is the general sense of the relation between humans and nature that it implies. Individuals with this perspective would obviously prefer an explanation for deformed frogs that makes no reference to human beings over an explanation that implicates them in the phenomenon. They would prefer a frog that, deformed or not, is just a frog, just a feature of the material world of nature, apparently independent of human life. But the idea of a frog as just a frog, with no nonmaterial link to human life—a purely objective frog—is really a historical invention. It is an invention that has proven to have great practical value, but is an invention nevertheless; which, like all cultural or personal possessions, can become an obstacle when held too dearly or uncritically.

This view of frogs, this view of nature, as essentially alien to humans—*it* as essentially matter, *we* as essentially mind or spirit—is readily evident among many promoters of free enterprise, largely because they unabashedly advocate the dominant ethical and economic system that is a logical counterpart of this epistemology. It is thus not surprising that they often find themselves at odds with environmentalists. But many of the skirmishes between these camps are rooted more in questions of practice than in contrasts of basic epistemology, since a great many members of the green community—sometimes unwittingly, unwillingly, and against their better judgment—also habitually experience nature as distinct from the defining essence of human life. The dualism of subject and object, the apparent independence of we as private selves "in here" from a world of public objects "out there," with the two realms communicating only through the senses, is a part of our cultural inheritance, and all of us, from across the political spectrum, have been in some measure shaped by this epistemological endowment.

It is typically considered simply common sense to see frogs, and nature as a whole, in this dualistic way, but we had to learn, both culturally and personally, to do so. It is difficult to see beyond common sense, even if the spirit is willing, but we can at least make a start by asking how this sense of things became so common. By doing so we will find that throughout much of Western history the "in here" and the "out there" were experienced as joined not just through the senses but also through a supersensible factor common to both, so that the "inner" life of the human self was felt as one with the life of the objects of nature. But we will also find that this dynamic polarity, this play of contrasting mental and material poles, was gradually fractured into the dualism with which we live.[3]

We can witness the emergence of the objectified frog by looking briefly at a few highlights of Western history from ancient Egypt to the eighteenth century. There is a great temptation to see the changing ideas of nature evident in this story as merely changing interpretations imposed by people of the past on a natural world that, at bottom, has always been populated by the same sort of objects we see in it today. But this would be to read our modern mindscape back into historically distant eras, before this mindscape had arisen. In the following sketch, I will resist this temptation and instead will follow the perhaps more demanding, but also more valid and interesting, path of seeing the changing perceptions of nature as indications of the historical evolution of the Western mind. And because our minds are our organs of perception, tracing the evolution of the mind enables us to watch the historical emergence of our familiar world. Evidence for this evolution can be found in many places in the historical record, but I will focus here on a small but pertinent element of it—the history of deformities.

Consider, first, how deformed frogs would most likely have been perceived in the ancient Near East. If such creatures had been discovered in the swamps of Egypt or Mesopotamia in the first millennium B.C.E., where the annual emergence of innumerable tadpoles and frogs with the yearly floods was a manifestation of nature's primordial creativity, they would surely have been taken quite seriously. We can imagine an ancient Ninevite, upon making such a discovery, scurrying to the vast omen literature in the library of King

Ashurbanipal, hoping for some hint of the future in the thousands of "if-then" statements carved in stone tablets. He would seek especially a relevant teratological omen, the special type dealing with the import of abnormal births or malformations in humans and animals. On one tablet, he might find help in the statement that "if an anomaly's eyes are brought together on its forehead—the land will be oppressed" or that "if an anomaly's eyes are on its back—the land will decrease in size." And the inscription on another that "if an anomaly's four legs turn toward the rear, and its eyes are on the right—whatever the enemy asks from you, give it to him willingly" might indeed afford the desired certainty.[4] Considering the great cosmological and agricultural significance associated with the lives of frogs, the prognosis probably would not have been bright. If a plague of frogs was an expression of divine disfavor with an Egyptian pharaoh, what would be the meaning of ill-formed frogs—another indication of divine displeasure, or perhaps a hint of divine impotence?

For these ancients, natural phenomena, especially exceptional ones, were practically vibrating with significance, their very bodies echoing with divine impulses and imperatives. They were of interest not so much as entities in themselves, however, but mostly as pointers to something else, as possible indicators of cosmic conditions and as clues about what to expect. If I, as an ancient Mesopotamian, were to find a deformed frog, I would take notice because I would experience it as issuing from a realm beyond itself and as having, because of that fact, reference beyond itself, reference to me and my household or tribe. The frog would not be what we today would call a biological specimen and would study to determine the cause or cure of its abnormality; it would be a communication, a sign, relying for its efficacy, like all communication, on some measure of commonality between the source and the sensitive receiver. The historical evidence indicates, moreover, that for the ancients, this was the natural way of seeing things. It was not that, as we might assume from our modern perspective, they saw a merely material frog and proceeded to attach all sorts of symbolic meanings to it. They apparently experienced the world in this way automatically, without much thought about it. Seeing a frog symbolically would have been "common sense" to them, just as seeing a frog nonsymbolically, we might say "literally," seems common sense to us.

If a misshapen frog had been scooped up a few centuries later by one of Aristotle's students on a walk through the garden of the Lyceum, the pioneering zoologist would have been most intrigued. He, in contrast to his earlier Near Eastern neighbors, would have been less prone to read divine imperatives in the frog and more likely to contemplate its place in the grand order of the cosmos. Given Aristotle's belief that "the craftsmanship of nature provides extraordinary pleasures for those who can recognize the causes in things,"[5] he and his followers would have considered the possible causes of the deformity.

And of these "causes," the kind most crucial for understanding biological processes, and the kind most controversial in the subsequent history of science, was the "final cause"—the goal toward which the process is directed, the fully formed frog that is the natural fulfillment of the egg's potential. The Aristotelians noticed that in the great majority of cases, organisms arrived successfully at their intended form, and they thus saw deformities as a failure of this formative pursuit. But it was a failure only in terms of nature's customary performance, not in terms of the multifarious designs of nature in toto.

For Aristotle, nature as a whole was a living organism that not only harbored the creative potential for a great variety of biological forms but also encompassed human life and thought. For him, the forces of nature were also the formative energies of the human mind. A tadpole's transformation into a frog involved the purposive realization of the egg's potential. Similarly, the simple act of beholding a frog, or of thinking a frog, involved the realization of the form of the frog in the person's mind. Seeing, hearing, touching, tasting, smelling the world thus implied a sort of identity or solidarity between the knower and the known, between the human mind and nature. In the very act of perceiving a deformed frog in an ancient Greek garden, then, Aristotle would have felt not only that he was witnessing a curious case of nature's formative work but also that he was participating in this universal process.

Aristotle's ideas about the generation of animals through the purposive "craftsmanship of nature" were elaborated with abandon in the Middle Ages and the Renaissance, as was his idea that gaining knowledge of nature involved not just a process of gathering data but an actual "in-forming" of the mind. For a great many medieval students of nature, frogs, or any other animal, existed primarily to teach people lessons, as is evident in myriad medieval bestiaries. In seeing natural things as whispered words of God, these medieval Europeans are somewhat reminiscent of the ancient Egyptians and Mesopotamians in their interpretation of omens. But the interpretation now, of course, had a Christian cast, with the Book of Nature, like the Book of Scripture, illuminating the heavenly path.

Some medieval naturalists, though, also found natural processes themselves quite interesting. If Albert the Great, that amazing thirteenth-century interpreter of Aristotle and teacher of Thomas Aquinas, had come across a deformed frog in the valley of the Rhine near his home, he would quite likely have looked to the hills and the heavens for answers. Albert, who was well acquainted with the metamorphosis of frogs, thought that, beyond the inner principles emphasized by Aristotle, various local geographical features as well as a rich array of immaterial influences from the stars were also implicated in biological processes. Although he believed that humans were perhaps more susceptible to astral influences than were animals, which he believed were more sensitive to atmospheric changes, he proposed that the conjunction of certain patterns of the planets with particular places on earth could prevent embryos from reaching their proper final form.[6] A deformed frog could thus be an indicator of prevailing cosmic dispositions, conditions also linked to the material and spiritual health of humankind. Throughout the Middle Ages, the Renaissance, and beyond, deformities were typically interpreted as some combination of failed final causes and unfriendly skies. Perhaps never, then, during this time was a frog just a frog; it was a sign from heaven, an image of a transcendent world that was also the sustaining leaven of human life.

But a sea change was on the horizon, a tidal shift in the history of Western consciousness, and the spirit that was wafting over European culture in the early seventeenth century is epitomized in the figure of Francis Bacon. For Bacon, the study of biological anomalies belonged wholly within the realm of *natural*, not *supernatural*, history. And Baconian natural history could never proceed so long as plants and animals were seen essentially as symbols enmeshed in a web of sympathetic relations, as divine utterances with depths of meaning. On the contrary, he argued that "the world is not the image of God," and that if we could clear the symbolic cobwebs from our minds and remove the scales of tradition from our eyes, the true and unequivocal language of nature would be plain for all to see. So Bacon, on the whole, vigorously rejected Aristotle's teleological philosophy of nature (though traces of final causes still crept into his thought) and the medieval emblematic universe it inspired. The Book of Nature no longer speaks, he said, so it is no use training one's ear to hear a still, small voice. Nature is silent, but clear, and requires simply a reliable reading.

Shorn of symbolism and moral implications, deformities could then fit quite well into Bacon's plan for the advancement of useful knowledge. For in the pursuit of such knowledge by "hounding nature in her wanderings," it was helpful to follow both the highways and the byways because "he who has learnt her deviations will be able more accurately to describe her paths." The more accurate and objective one's knowledge of nature, then, the more successful one could be in applying this knowledge "to the effecting of all things possible" for the betterment of humankind. There were two aspects to this Promethean project, and their linkage is evident in the famous Baconian adage that "knowledge is power." There is the epistemological aspect—that humankind needed a new way of knowing nature—and there is the ethical aspect—that humankind desired a new way of using nature. And these aspects go together like hand and glove.[7]

We thus see emerging a relatively new species of frog—a frog that is just a frog, bearing no transcendental stamp or moral message, a merely material frog. And it seems clear that such a frog, that such a natural world, was linked historically to the very deliberate invention of the new scientific method for knowing and using nature. It is a method that took pains to separate the material world "out there" from the human mind "in here," a perspective that dissolves the solidarity of humanity and nature and stands in such contrast to the ancient and medieval sense that knowledge involves nature re-forming herself in the mind. This method was, of course, developed even more rigorously by Bacon's more mathematical cohorts, Galileo and Descartes, who proposed that nature consists, in essence, of merely what is measurable. Curiously, for these seminal architects of objectivity, even the color, sound, smell, taste, and feel of a frog exist merely in the mind of the perceiver; only the animal's size, shape, weight, and movement persist after the naturalist quits the marshland and goes indoors.

But the objectified frog, at this point, was still but partly liberated from the waters of symbolism. Its full emergence into the modern light of day, so that almost anybody could get a good look, would require much further expansion and assimilation of the new scientific worldview. Bacon's spirit of observation and experimentation stimulated later naturalists such as Thomas Browne and John Ray and was infused into the communal body of the world's first major scientific society, the Royal Society of London, founded in 1660. The Royal Society displayed a lively interest in biological anomalies, but it was the Academy of Sciences in Paris, in concert with the Museum of Natural History, that gave to biological specimens their much more modern feel. Here, from the late seventeenth century, scientific specialists investigated a great many deformities from the perspective of disciplines such as embryology and comparative anatomy. This systematic medicalization of anomalies furthered the movement toward thoroughly naturalistic explanations.[8] The objectified frog was clambering onto the shore. And over the next two centuries, through innumerable field investigations, laboratory studies, and educational endeavors, this new species of frog, by following this naturalistic path, came to rest in full view of professionals and the general public alike. The evolution of the Western mind—the evolution of our familiar world—had achieved one of the most distinctive features of its modern form.

Objectified frogs, whether malformed or whole, can of course be easily eaten, dissected, or ignored with impunity. Aside from the nourishment they have provided through the centuries, especially to the French, they have afforded much valuable biological and medical knowledge, which most likely would not have occurred had they, and nature as a whole, remained immersed in symbolism. The labored invention of the objectified frog, of an objectified natural world, was in essence a positive development. It hardly requires saying that this way of being in the world, having launched the whole grand enterprise of modern science and technology, has had undeniable benefits. But it is also rather painfully obvious, now that the project has played itself out on the planet for four centuries, that it has its limitations.

The main drawback of this mode of perception is its tendency to fragment the world, typically dividing the phenomena of nature into relatively manageable provinces. This is an understandable strategy, of course, but it can lead to some unfortunate provincialism. A six-legged frog would traditionally have been a fascination to anatomists, physiologists, and embryologists. But hydrologists, botanists, and climatologists, not to mention foresters and engineers, probably would have paid little notice. The modern scientific project, as we have seen, also sets up sovereign realms of matter and mind, a divorce that has given many fine philosophical minds epistemological fits and has left even many ardent nature lovers with but faint hope of a thoroughgoing human-nature reunion.

The science of ecology, with its emphasis on relationships, grew out of the recognition that a fragmented science cannot do justice to an actually integrated world. Ernst Haeckel, inspired by Charles Darwin, coined the term for the new field in 1866 to designate the holistic, and scientifically rigorous, effort to understand an organism not in isolation but in relation to the larger household of nature, and ecology thus draws on diverse scientific subfields. Ecologists came to see human beings, too, as elements of ecosystems, a recognition that gave birth, of course, to the modern environmental movement. Now, ecological *science* can get on relatively well without worrying much about that other fragmentation of the world built into the Baconian program, the division between the sentient human mind and the insensate material world. But the broader ecological *sensibility* of environmentalism, with its larger ethical dimension, should, and sometimes does, find this yawning gulf more of an obstacle. In the early 1920s, Aldo Leopold was already realizing that a scientific—even an ecological—understanding of nature was not a firm enough foundation upon which to build a conservation ethic. He remarked that:

> Possibly, in our intuitive perceptions, which may be truer than our science and less impeded by words than our philosophies, we realize the indivisibility of the earth—its soil, mountains, rivers, forests, climate, plants, and animals, and respect it collectively not only as a useful servant but as a living being, vastly less alive than ourselves in degree, but vastly greater than ourselves in time and space—a being that was old when the morning stars sang together, and, when the last of us has been gathered unto his fathers, will still be young.[9]

If Baconian "knowledge" is tied to power over nature, then Leopoldian "intuitive perception," it seems, is linked to respectful citizenship within the biotic community.

What truly hinders this sort of responsible citizenship is that for so many of us today, the "scientific," objective view of nature feels so thoroughly natural, so "intuitive," because we engage in it so automatically, without thinking about it. Just as the ancient Mesopotamians would have naturally experienced a deformed frog as a divine omen, and twelfth-century Europeans might have naturally seen one as a sign from the heavens, so is it second nature for us to see such a frog as just a frog, with no particular reference to us. But such a frog would never have appeared, and would not appear to us now, had the story I have sketched in this essay not occurred and woven its way into the personal history of each one of us. Our world is populated with objectified frogs because of our objectifying minds. Thus, they all have reference to us—if not always to our way of living, inevitably to our way of knowing.

An interesting bit of perceptual entertainment called the "magic eye" has gained enormous popularity in recent years. It involves holding an apparently nonsensical pattern close to one's eyes, softening one's focus, and moving the pattern away until—usually after some effort and experimentation—familiar three-dimensional objects stand out rather surprisingly from the page. The objects are usually a little tenuous at first; it is easy to lose them if one's focus or perspective changes. But they reappear more readily with repeated viewing. All of which makes it obvious that the objects would not appear at all if one were not looking in the proper way, and that they would sink again into the background if one closed one's eyes.

An incredibly rich world of three-dimensional objects answers to our opening eyes each morning and meets us at every turn. We are so practiced in this manner of seeing that we are oblivious to the effort and experimentation that were required in both our cultural and our personal history to learn to see in this way. We see the world with our magic eyes. And it refers to us by its very nature.

The vigor with which latter-day Baconians have appropriated some apparent evidence that deformed frogs might have no reference to humankind clearly reveals how the phenomenon of the objectified frog is still so closely linked to the Promethean program of industrial growth. The degree to which humankind is ethically implicated in these frogs is yet to be determined. But the appearance of any merely material frog on the shores of our perceptual worlds always implicates us epistemologically. And our knowing the world in this way, as essentially a material resource, is the license that, when joined with desire, has unleashed untold exploitation—of land, and water, and the habitats of frogs.

The great virtue of ecological *science* lies in its efforts to soften the fragmentizing tendencies of modern science and to see the world whole. With this perspective in mind, even schoolchildren realize that because we share a planet with innumerable other organisms, deformed frogs in a Minnesota field just might have reference to our behavior. But a deeper ecological *sensibility* should also encompass our way of knowing. As Aldo Leopold noticed in the early days of environmentalism, the best foundation for environmental ethics lies in the sense of "a closer and deeper relation" between humans and nature "than would necessarily follow the mechanistic conception of the earth as our physical provider and abiding place."[10] It is possible that we share with the frogs not only a planet but also, at least in some small measure, an essence.

When the early modern philosophers established the great divide between mind and nature, between subject and object, all human beings were, at least theoretically, included on the "subject" side of the line and thus were seen as worthy of ethical consideration. In practice, of course, it has not always worked that way; innumerable individuals throughout history have been treated more as objects than as true subjects. There seems to be no end to the mischief caused by the refusal to grant subjecthood to other human beings.

Or to other nonhuman beings. Suppose that, while not sacrificing the rigorous pursuit of detailed knowledge that vitalizes modern science, we were to recover some sense of nature's depth that was submerged in the scientific wake. Suppose we agreed to bridge the great divide and see nature as somewhat more like us—with some measure of an integrated inner life—than like a collection of mere objects sans *psyche*. A frog with two legs or four would then merit our careful consideration, not only for what it might mean for us but also for what it means to itself. If it were possible for us to consider a frog from the inside—not just in terms of the human project but in terms of nature's project, of which we also are an outcome—we could divine more than just oracles for our way of living. We could discover the ground for a deeper relation in the amphibian's revelations of life on the edge—revelations of life at the boundary of the wet and the dry, in sympathy with the seasons, at the primordial threshold of silence and voice; maybe even of life at the crossing of nature and culture, and of the inherently precarious business of living on both sides.

Figure 15
Missing link. (After *The Riverside Natural History,* 1888, facing p. 337.)

Epilogue
A Matter of Frogs: Children of Mountain and Lake
(with Apologies to Aristophanes)

Almost forty years ago, when our family of five acquired forty acres in the Front Range west of Colorado Springs, one of the pleasures for this city girl was the leopard frogs. I knew what frogs were—amphibians—but that were about all. I delighted, those first summers, in walking around the pond and hearing a juicy *plop* every few feet or so, or seeing a whiz of green and a snap of white as the leopard frogs flew into the water, forming "stylish triangles with their ballet dancer's legs," in the words of Scottish poet Norman MacCaig.

I have a photograph of one, secure and sedate under an inch of water, undoubtedly feeling safe because it had the ceiling on its back. About four inches long, slender, it bore beautiful dark spots outlined in pale yellow, arranged elegantly on either side of two yellowish stripes beginning at the eyes and running parallel down its back. Dark, pale-edged stripes banded its legs. The cryptic coloring of the leopard frog matches its habitat: it frequents pond edges and meadows, and its markings provide good camouflage; the dark spots may also provide some thermal benefit on cold mountain mornings. Back then, though, even though I knew that this type of frog was the most common and widespread in North America, as witnessed by its several common names, this quiet soul gleaming beneath the water was *my* frog in *my* pond, and therefore special to me.

That perusal came in mid-June, according to my journal notes, and I'd already found some skeins of eggs in the sedges and grasses that fringed the pond. Earlier, I'd heard the frogs' call, a guttural, unharmonious sound followed by some clucks, not the most euphonious call among frogs. Robert Froman, in *Spiders, Snakes, and Other Outcasts,* describes the difference between the calls of spring peepers, "which have high sweet trilling voices," and those of the "decidedly unmusical leopard frogs and wood-frogs, which sound like a distant procession of ox-drawn wagons, the axles of which badly need lubrication." Another writer describes them as the sound made by rubbing your fingers over an overinflated balloon. I thought they were beautiful.

Those first frog years marked the beginning of my understanding many things about this mountain land, and I remember them clearly. I also learned that beginnings are noteworthy, but endings are amorphous. We acknowledge, often celebrate, beginnings, raising a glass of champagne or noting them in a journal. But it is not so with endings because we are seldom sure when they occur. Was this really the last frost? Or will there be others? And half past summer, we wake up and realize that the mornings have all turned warm and we can't remember when that last frost really did occur—we just know that now it's summer and things are different.

A lot of years passed before I woke up to the fact that nothing was jumping into the lake when I walked around it. Somewhere between an amorphous then and a definite now, the frogs had disappeared. Where were they? Where, indeed, were the frogs—for about that time, I began reading about the dearth or absence of frogs all over the world and various suggestions as to what might have happened to them, all of which were worrisome.

Frogs may have played a limited role in my life, but it has been a decisive one. I first became aware in my late twenties, for heaven's sake, that frogs actually sang, and that was coupled with a wonderful moment. I had just returned to Massachusetts, where I was teaching; it was early evening, and my fiancé had met me at the train station. As we drove north, the pulsing trills of hundreds of frogs filled the air with sound. I had no idea that these were the voices of frogs, but a native soon explained that they were spring peepers, *Hyla crucifer,* the darlings of New England springs, and that their scientific name derived from the pale cross marking their backs. To be serenaded on my return was a romantic moment, something I have never forgotten, along with being bowled over by the decibel level of their sound. Spring peepers put out levels of sound far beyond their size, as do all frogs bound to temporary waters—a fact I discovered some forty years later when listening to the Gatling gun blasts of Pacific tree frogs deep in the Grand Canyon. Being small, spring peepers also sing in a higher voice than, say, bullfrogs, with their thumping *kerr-ump.*

Then there were the charming, tiny frogs on the Juan Fernández Islands, four hundred miles west of Santiago, Chile, literally in the middle of nowhere. I had gone there on assignment to be a latter-day Robinson Crusoe (one of the islands in the archipelago, Más a Tierra, is also known as Robinson Crusoe Island). There, nestled under a watercress leaf beside a twinkling rivulet in the grass, crouched an adorable frog that could have fit on a quarter. (One of the puzzles of island biogeography is how animals got to such faraway, isolated places; in this case, frog eggs may well have come in on the feet of birds.) The delicate, amorous warblings of these tiny frogs gentled my evenings spent alone in Crusoe's cave, and I wondered whether they had been here when Alexander

Selkirk, the model for Daniel Defoe's Crusoe, lived out his four years on this island. Had their delicate, tinkling voices woven through his falling sleep as they wove through mine?

In between these encounters, I have admired pictures of tropical tree frogs, flagrantly colored outrageous blues and yellows and greens, and have come to admire frogs' adaptability and to love their songs, especially when each solitary singer joins into a chorus of slippery invitations.

So when the frogs' populations declined precipitously, it was of personal concern. For me, it's a philosophical question. Do humans have the right to change environments so that other creatures can no longer live there? Our world society, if it thinks about the matter at all, generally seems to consider that the only way for the human race to live is by the destruction of others' habitats to ensure enough space for itself. A great many people react quite belligerently to any suggestion that small, slippery creatures that don't vote and whose voices don't make a lot of sense to anyone but another frog are worthy of a change in our plans. I grew up in a time when that viewpoint was so prevalent that no one even thought about it; there was plenty of space, plenty of clean water, and plenty of frogs. Now I live in a time when that approach no longer makes sense. We have come to recognize that the structural vitality of our world depends on diverse populations of creatures living out their lives, tiny and large, in the habitats to which they have worked long and hard to adapt over the millennia. We do comprehend that all these crosscurrent lives make a thick, sturdy fabric, and that when we start pulling out threads and the fabric thins, the possibilities are no more enticing than wearing clothes so threadbare that they no longer protect our tender bodies. I wish it weren't so, and I wish we hadn't banged our noses against the inevitability of nature; I wish there were still a plethora of clean water and air and that we had not done anything to diminish an environment that supports and nourishes us all. I don't want to live in a world of only crows and cockroaches, cheatgrass and leafy spurge, sparrows and starlings. There's something about being human that gives me joy in beautiful colors and enticing songs and different shapes, that drives me to learn about creatures with four legs, six legs, and dozens of legs.

In my lifetime, the threads have grown thinner. Some of them have already frayed and left holes: we no longer have carrier pigeons or dodos. Some of the threads have broken for certain groups, as they have for some amphibians, especially those with thin, moist skins, dependence on water, feet with nuptial pads, and goggly eyes. The disappearance of frogs has come about in such a short time span that it takes my breath away. To lose a frog is to lose the good health of the world.

I keep hoping that some bird (but they, too, are less numerous where we live) will bring back some frog eggs stuck to its feet and rinse them off in our pond, and next spring I'll walk around the lake and hear a suggestive croak or see a sleek, green body, legs stretched out behind, clear a grass stem and prong into the water. Because I'm an optimist, I'll keep hoping. I miss them very much. In the place I love most in all the world, something is missing, something is awry. There are no frogs here, where they belong.

I commend leopard frogs to be quick and proficient about it: Lay thousands of eggs where a bird may step, bless them in their translucent purses of gelatin, let your good wishes prime their tiny spirits, round their toes, brighten their eyes, inscribe bright rings of handsome color on their backs. Make medicine bundles for them as I shall, and send them back to this pond—such a small thing—a frog's pulsing message, a minute plash at a pond's edge, but it signifies so much. Please arrange to ship your dark dots of encapsulated progeny through the air and back here, to hatch into sleek leapers with their song that will weave me back into their wiggly, slippery lives.

—ANN HAYMOND ZWINGER

Notes

Introduction
1. T. Browne, *Religio Medici* (London: Macmillan, 1926), 55.
2. A. O. Lovejoy, *The Great Chain of Being* (Cambridge, Mass.: Harvard University Press, 1936; 1964), 16.
3. H. More, *An Antidote Against Atheism* (London: Roger Daniel, 1653), 81.
4. T. Browne, *Religio Medici* (Edinburgh: John Grant, 1912), 26.
5. C. Linnaeus, in *A Trilogy on the Herpetology of Linnaeus's "Systema Naturae" X,* edited by K. Kitchell Jr. and H. A. Dundee (Washington, D.C.: Smithsonian Institution Press, 1994), 5–6.
6. T. Sprat, *History of the Royal Society* (London: Martyn, 1667), 110.
7. T. H. Huxley, *Evolution and Ethics* (Princeton, N.J.: Princeton University Press, 1989), 77.
8. Aristophanes, *The Frogs* (London: J. M. Dent; New York: E. P. Dutton, n.d.), 15.

Part I: Interpreting the Cosmos
1. Aristotle, *Parts of Animals,* translated by A. L. Peck. Loeb Classical Library (Cambridge, Mass.: Harvard University Press, 1937), I.v. (645a 24).
2. Galileo, *The Assayer,* in *Discoveries and Opinions of Galileo,* translated by S. Drake (New York: Anchor, 1957), 238.
3. F. Jacob, *The Logic of Life* (New York: Random House, 1973), 28–29.

Part II: Reclaiming Paradise
1. J. Ray, *The Wisdom of God Manifested in the Works of the Creation* (London: Benjamin Walford, 1691), 165.
2. J. Burroughs, "Gilbert White's Book," in *Indoor Studies* (Boston: Houghton Mifflin Company, 1889, 1895, 1904), 179.
3. H. D. Thoreau, *Journal* entry for February 18, 1860, in *The Writings of Henry David Thoreau* (Boston: Houghton Mifflin Company, 1906).
4. J. Burroughs, "The Gospel of Nature," in *Time and Change* (Boston: Houghton Mifflin Company, 1913), 250–251.

Part III: Telling Naturalistic Tales

1. C. Darwin, quoted in D. Worster, *Nature's Economy*, 2nd ed. (Cambridge: Cambridge University Press, 1994), 128.
2. W. Irvine, *Apes, Angels, and Victorians* (New York: McGraw-Hill, 1955), 313.

Stephen Jay Gould, "Here Goes Nothing"

1. C. J. Corben, G. J. Ingram, and M. J. Tyler, "Gastric Brooding: Unique Form of Parental Care in an Australian Frog," *Science* 186 (6 Dec. 1974): 946–947.
2. M. J. Tyler, ed., *The Gastric Brooding Frog* (London and Canberra: Croom Helm, 1983).
3. R. W. McDiarmid, "Evolution of Parental Care in Frogs," in *The Development of Behavior: Comparative and Evolutionary Aspects*, edited by G. M. Burghardt and M. Bekoff, 127–147 (New York: Garland, 1978).
4. G. J. Ingram, M. Anstis, and C. J. Corben, "Observations on the Australian Leptodactylid Frog, *Assa darlingtoni*," *Herpetologia* 31 (1975): 425–429.
5. K. Busse, "Care of the Young by Male *Rhinoderma darwini*," *Copeia*, No. 2 (1970): 395.

Part IV: Remembering the Earth

1. J. W. Krutch, *The Best Nature Writing of Joseph Wood Krutch* (New York: Morrow, 1969), 13–14.

Part V: Reading the Signs of the Times

Gordon L. Miller, "Dimensions of Deformity"

1. See S. K. Sessions, R. A. Franssen, and V. L. Horner, "Morphological Clues from Multilegged Frogs: Are Retinoids to Blame?" *Science* 284 (30 April 1999): 800–802; P. T. J. Johnson et al., "The Effect of Trematode Infection on Amphibian Limb Development and Survivorship," *Science* 284 (30 April 1999): 802–804.
2. L. H. Rockwell, "Jumping to Conclusions," *Journal of Commerce* (5 May 1999): 5A; see also M. Fumento, "With Frog Scare Debunked, It Isn't Easy Being Green," *Wall Street Journal* (12 May 1999): A22; B. Doherty, "Amphibian Warfare," *Weekly Standard* (24 May 1999): 16–18. (Mr. Rockwell is president of the Ludwig von Mises Institute; Mr. Fumento is a fellow at the Hudson Institute; and Mr. Doherty is a fellow at the Competitive Enterprise Institute.)
3. I am greatly indebted to thinkers such as Owen Barfield, particularly to his *Saving the Appearances* (New York: Harcourt, Brace & World, 1965), and Theodore Roszak, especially his *Where the Wasteland Ends* (New York: Doubleday, 1972), for insightful explorations of the issues. I have also borrowed from Roszak the term "mindscape" in the paragraph that follows.
4. E. Leichty, *The Omen Series Šumma Izbu* (Locust Valley, N.Y.: J. J. Augustin, 1970), 125, 126, 157.
5. Aristotle, *Parts of Animals*, 645a7f. Quoted in G. E. R. Lloyd, *Early Greek Science: Thales to Aristotle* (New York: W. W. Norton & Company, 1970), 105.

6. Albert the Great, *Man and the Beasts,* books 22–26 (Binghamton, N.Y.: Medieval and Renaissance Texts and Studies, 1987), 440; see also C. J. Glacken, *Traces on the Rhodian Shore* (Berkeley: University of California Press, 1967), 227–229, 265–271.
7. F. Bacon, *Advancement of Learning, Novum Organum, New Atlantis* (Chicago: Encyclopaedia Britannica, 1952), 33, 159, 210.
8. K. Park and L. J. Daston, "Unnatural Conceptions: The Study of Monsters in Sixteenth- and Seventeenth-Century France and England," *Past and Present* 92 (August 1981): 51–53.
9. A. Leopold, "Some Fundamentals of Conservation in the Southwest" (1923), *Environmental Ethics* 1 (summer 1979): 140.
10. Ibid., 139.

Sources

The text selections in this book were drawn from the following sources. I am grateful to the publishers and authors indicated for permission to reprint copyrighted material.

Part I: Interpreting the Cosmos

Aristotle. *Historia Animalium (Inquiry Concerning Animals)*. From *The Works of Aristotle*, translated under the editorship of J. A. Smith and W. D. Ross. Vol. 4, by D. W. Thompson. Copyright © 1910. Oxford, England: Clarendon Press.

Pliny the Elder. Reprinted by permission of the publishers of the Loeb Classical Library from Pliny, *Natural History*. Vol. 3, translated by H. Rackham (Cambridge, Mass.: Harvard University Press, 1940). Vol. 8, translated by W. H. S. Jones (Cambridge, Mass.: Harvard University Press, 1963).

Anonymous. *Physiologus*. Translated by James Carlill. From *The Epic of the Beast*. Copyright © 1924. London: George Routledge & Sons; New York: E. P. Dutton.

Albert the Great. *De Animalibus (Man and the Beasts)*. Translated by James J. Scanlan. Copyright © 1987. Binghamton, N.Y.: Medieval and Renaissance Texts and Studies. Reprinted by permission of the Center for Medieval and Renaissance Studies.

Edward Topsell. *The History of Four-Footed Beasts and Serpents*. London: printed by E. Cotes for G. Sawbridge, T. Williams, and T. Johnson, 1658.

Thomas Browne. *Pseudodoxia Epidemica*. From *The Works of Sir Thomas Browne*, edited by Charles Sayle. Copyright © 1912. Edinburgh: John Grant.

Jan Swammerdam. *The Book of Nature; or, The History of Insects*. Translated by T. Flloyd; revised and improved by J. Hill. London: printed for C. G. Seyffert, 1758.

John Ray. *The Wisdom of God Manifested in the Works of the Creation*. 5th ed. London: printed by J. B. for B. Walford, 1709.

Part II: Reclaiming Paradise

Gilbert White. *The Natural History of Selborne*. Copyright © 1924. London: Arrowsmith.

William Bartram. *Travels Through North and South Carolina, Georgia, East and West Florida*. Philadelphia: printed by James and Johnson, 1791.

Henry David Thoreau. *Journal*. From *The Writings of Henry David Thoreau*. Copyright © 1906. Boston and New York: Houghton Mifflin Company.

John Burroughs. "The Tree-Toad." From *The Writings of John Burroughs*. Vol. 5, *Pepacton*. Copyright © 1904. Boston and New York: Houghton Mifflin Company.

W. H. Hudson. "The Toad as Traveller." From *The Book of a Naturalist*. Copyright © 1919. New York: G. H. Doran.

Part III: Telling Naturalistic Tales

Charles Darwin. *The Voyage of the Beagle*. Copyright © 1909. New York: P. F. Collier & Son. *The Origin of Species*. Copyright © 1898. New York: D. Appleton and Company.

Thomas Henry Huxley. "On the Hypothesis That Animals Are Automata, and Its History." From *Science and Culture*. Copyright © 1893. New York: D. Appleton and Company.

J. Arthur Thomson. "The Tale of Tadpoles." From *The Biology of the Seasons*. Copyright © 1911. London: A. Melrose.

Julian Huxley. "The Frog and Biology." From *Essays in Popular Science*. Copyright © 1927. New York: Alfred A. Knopf.

Loren Eiseley. "The Dance of the Frogs." From *The Star Thrower*, by Loren Eiseley. Copyright © 1978 by The Estate of Loren C. Eiseley, Mabel L. Eiseley, Executrix. Reprinted by permission of Times Books, a division of Random House, Inc.

Stephen Jay Gould. "Here Goes Nothing." From *Bully for Brontosaurus: Reflections in Natural History*, by Stephen Jay Gould. Copyright © 1991 by Stephen Jay Gould. Reprinted by permission of W. W. Norton & Company, Inc.

David Scott. "A Breeding Congress." Reprinted with permission from *Natural History* (October 1998). Copyright the American Museum of Natural History (1998).

Part IV: Remembering the Earth

Dallas Lore Sharp. *The Face of the Fields*. Copyright © 1911. Boston and New York: Houghton Mifflin Company. "My Twenty-Four-Dollar Toad." From *Sanctuary! Sanctuary!* Copyright © 1926. New York and London: Harper & Brothers.

Donald Culross Peattie. *An Almanac for Moderns*. Copyright © 1935 by Donald Culross Peattie renewed. Reprinted by permission of Curtis Brown Ltd. and Noel Peattie. New York: G. P. Putnam's Sons.

George Orwell. "Thoughts on the Common Toad." From *The New Republic* (May 20, 1946).

Joseph Wood Krutch. "The Day of the Peepers." From *The Best Nature Writing of Joseph Wood Krutch*. Copyright © 1949–1970 Joseph Wood Krutch. Reprinted by permission of HarperCollins Publishers, Inc.

Edwin Way Teale. "Audubon's Salamanders." Copyright © 1965 by the estate of Edwin Way Teale. From *Wandering Through Winter*. Reprinted by permission of St. Martin's Press, LLC.

Annie Dillard. *Pilgrim at Tinker Creek*. Copyright © 1974 by Annie Dillard. Reprinted by permission of HarperCollins Publishers, Inc.

Robert Michael Pyle. *Wintergreen: Listening to the Land's Heart*. Copyright © 1986 by Robert Michael Pyle. Boston: Houghton Mifflin Company. Reprinted by permission of the author.

Ann Haymond Zwinger. *The Mysterious Lands: A Naturalist Explores the Four Great Deserts of the Southwest*. Copyright © 1989 by Ann Haymond Zwinger. New York: E. P. Dutton. Reprinted by permission of the author. *Downcanyon: A Naturalist Explores the Colorado River Through the Grand Canyon*, by Ann Haymond Zwinger. Copyright © 1995 by Ann Haymond Zwinger. Reprinted by permission of the University of Arizona Press and the author.

Terry Tempest Williams. *Desert Quartet*, by Terry Tempest Williams. Copyright © 1995 by Terry Tempest Williams. Reprinted by permission of Pantheon Books, a division of Random House, Inc.

Part V: Reading the Signs of the Times

Kathryn Phillips. *Tracking the Vanishing Frogs: An Ecological Mystery*, by Kathryn Phillips. Copyright © 1994 by Kathryn Phillips. Reprinted by permission of St. Martin's Press, LLC.

Laurie J. Vitt, Janalee P. Caldwell, Henry M. Wilbur, and David C. Smith. "Amphibians as Harbingers of Decay." From *BioScience* 40, no. 6 (1990). Reprinted by permission of the authors.

Andrew R. Blaustein. "Amphibians in a Bad Light." Reprinted with permission from *Natural History* (October 1994). Copyright the American Museum of Natural History (1994).

Timothy R. Halliday and W. Ronald Heyer. "The Case of the Vanishing Frogs." From *Technology Review* (May–June 1997). Reprinted by permission of the publisher.

Stephen Leahy. "The Sound of Silence." From *Nature Canada* (autumn 1998). Reprinted by permission of the author.

Virginia Morell. "Are Pathogens Felling Frogs?" Reprinted with permission from *Science* 284 (30 April 1999). Copyright © 1999 American Association for the Advancement of Science.

Jocelyn Kaiser. "A Trematode Parasite Causes Some Frog Deformities." Reprinted with permission from *Science* 284 (30 April 1999). Copyright © 1999 American Association for the Advancement of Science.

Acknowledgments

It is a pleasure to express my gratitude to the people who contributed to the realization of this book: To Bob Pyle and Ann Zwinger, who showed early enthusiasm for the project and graciously agreed to be the book's alpha and omega. To Barbara Dean and Barbara Youngblood at Island Press, who were wonderfully responsive, skillful, and encouraging editors, to Pat Harris for her superb copyediting, and to Cecilia González for her careful shepherding of the book through the production process. To Dave Madsen of Seattle University, for the translating of Latin and Greek phrases in parts I and II. To Jill Moerk, Karen Gilles, and other members of the Lemieux Library faculty and staff at Seattle University, who provided much expert and efficient bibliographic help. To Stan Sessions of Hartwick College, who gladly furnished an original image of a deformed frog. To Kevin Dann and María Bullón-Fernández, fine scholars and good friends, who offered some most helpful and perceptive suggestions about portions of the text. And to Jacquelyn Miller, who provided valuable editorial insights and, as always, indispensable practical and personal support.

About the Authors

ALBERT THE GREAT was a thirteenth-century Dominican monk who taught at the University of Paris, where Thomas Aquinas was his starring student, traveled widely in his native Germany and elsewhere as a provincial bishop, and was an innovative naturalist and a major force in introducing Aristotle to medieval Europe. Albert was declared a saint in 1931.

ANONYMOUS. *Physiologus* is a term that originally referred to the author of a second-century book devoted to interpreting the transcendental significance of the natural world. However, because the identity of this author was lost in the mists of history, the term has come to signify the book itself.

ARISTOTLE was a Greek natural philosopher of the fourth century B.C.E. He attended Plato's Academy for many years but eventually founded his own school, the Lyceum, and developed a distinctive understanding of the natural world that dominated European science and cosmology until the seventeenth century.

WILLIAM BARTRAM, born of Quaker parents in Philadelphia in 1739, was the first accomplished American artist-naturalist. The illustrations and descriptions in his *Travels* introduced many new species into the zoological canon, including many of the reptiles and amphibians of Florida, and delighted European readers eager for firsthand accounts of the American wilderness.

ANDREW R. BLAUSTEIN, a professor of zoology at Oregon State University, is the leading researcher of the link between ozone depletion and amphibian decline. He is a member of the Species Survival Commission of the World Conservation Union and has served as cochairman of the Pacific Northwest Section of the Declining Amphibian Populations Task Force.

THOMAS BROWNE was a seventeenth-century English author and physician known for his affecting literary expressions of his Christian faith. In his influential *Religio Medici,* written around 1635, he offers an integrated vision of religion and science. He was knighted in 1671.

JOHN BURROUGHS was one of the most popular American nature writers of the late nineteenth and early twentieth centuries. Anchored to his farm in the Hudson Valley, he cultivated fruit trees, friendships with leading lights such as Walt Whitman, John Muir, Theodore Roosevelt, Thomas Edison, and Henry Ford, and a simple literary style that effectively related his rambles in the countryside and the insights he gleaned from evolutionary biology.

JANALEE P. CALDWELL is associate curator of amphibians at the Sam Noble Oklahoma Museum of Natural History and an associate professor of zoology at the University of Oklahoma.

CHARLES DARWIN, grandson of another prominent evolutionist, Erasmus Darwin, was born in England in 1809. He undertook medical training at the University of Edinburgh and ministerial studies at Cambridge University but pursued his passion for natural history, which led him on board the HMS *Beagle* and eventually to his revolutionary theories. In addition to *The Origin of Species* (1859) and *The Descent of Man* (1871), and in spite of persistent ill health, he wrote voluminously, and in a pleasing style, on a great range of subjects until his death, in 1882. He is buried in Westminster Abbey, near Isaac Newton.

ANNIE DILLARD grew up in Pittsburgh in the 1950s and attended Hollins College, near Roanoke, Virginia. Her writing conveys a vivid sense of the drama and mysteries of life and an ardent invitation to launch into the deep. *Pilgrim at Tinker Creek,* her Thoreauvian "meteorological journal of the mind" set in the Blue Ridge of Virginia, won the Pulitzer Prize for nonfiction in 1975. Dillard is also the author of, among other works, *Teaching a Stone to Talk* (1982), *An American Childhood* (1987), and *For the Time Being* (1999). She has lived in Middletown, Connecticut, for many years and has taught at Wesleyan University.

LOREN EISELEY, a native of Nebraska, spent most of his professional life at the University of Pennsylvania, where he was a professor of anthropology and the history of science until his death, in 1977. He is the author of such widely read books as *The Immense Journey* (1957), *Darwin's Century* (1958), and *The Unexpected Universe* (1969). His writing typically combines natural history with autobiography, points to the limits of pure reason, and highlights occasions of special communion with the natural world.

STEPHEN JAY GOULD teaches geology, biology, and the history of science at Harvard University. With Niles Eldredge, he formulated the evolutionary theory known as "punctuated equilibrium." For twenty-five years, he has written the Darwin-inspired and historically informed "This View of Life" column in *Natural History Magazine,* producing essays that now fill several very popular books. He is also the author of *The Mismeasure of Man* (1981) and the best-selling *Wonderful Life* (1989), among other works.

TIMOTHY R. HALLIDAY is the international director of the Declining Amphibian Populations Task Force and works out of the Biodiversity Programs Office of the Smithsonian Institution.

W. RONALD HEYER is chair of the Declining Amphibian Populations Task Force, based at the Open University in Milton Keynes, England.

W. H. HUDSON was born in Argentina of American parents in 1841 and grew up there on the family farm. He moved to London in 1874 and wrote many books about his personal acquaintances with animals in both South America and England. His best-known work is the 1904 *Green Mansions,* a romantic fantasy with an elusive, birdlike heroine set in the mysterious jungles of Venezuela.

JULIAN HUXLEY was an English biologist born in 1887, during a time when scientific thought was leading to what he later described as the "eclipse of Darwinism." He was instrumental in revitalizing Darwinian theory by combining it with genetics. Intent on expanding Darwinism into a comprehensive worldview, he wrote numerous general books on this and other topics, as well as editing the diary of his grandfather, Thomas Henry Huxley, from his voyage on the HMS *Rattlesnake*. He held several prominent positions, including director general of the United Nations Educational, Scientific, and Cultural Organization (1946–1948). He died in 1975.

THOMAS HENRY HUXLEY was a nineteenth-century English biologist who was a close friend of Darwin and a staunch advocate of Darwinism. His voyage around Australia aboard the HMS *Rattlesnake* helped establish his scientific reputation, just as the voyage of the HMS *Beagle* had done for Darwin. In the decades following his voyage, he exerted great influence on public opinion through books such as *Man's Place in Nature* (1864) and numerous essays on biology and philosophy, which were informed by an agnosticism rooted in his refusal to venture beyond the evidence of the senses or the dictates of logic.

JOCELYN KAISER is a writer for *Science.*

JOSEPH WOOD KRUTCH turned to nature writing after an accomplished career as a drama critic for the *Nation* and as a professor of literature at Columbia

University. Born in Knoxville, Tennessee, in 1893, he lived for many years in New York and Connecticut but moved to Arizona in the late 1940s to consider more closely the patterns of nature and their import for human life. His numerous books include *Henry David Thoreau* (1948), *The Twelve Seasons* (1949), *The Voice of the Desert* (1955), and *The Great Chain of Life* (1956).

STEPHEN LEAHY lives in Brooklin, Ontario, where he is a writer and editor and contributes to magazines such as *Nature Canada, Equinox,* and *Audubon.*

GORDON L. MILLER is a native of Indiana, where the Hoosier ponds and streams fostered his lifelong fascination with amphibians. He attended Milligan College, Rutgers University, and Cambridge University and teaches in the history department and the Ecological Studies Program at Seattle University. His books include *Thirty Walks in New Jersey* (1992), with Kevin Dann, and *Wisdom of the Earth* (1997). He is currently writing a book about frogs.

VIRGINIA MORELL, author of *Ancestral Passions* (1995), a Leakey family biography, is a correspondent for *Science* and a contributing editor for *Discover.*

GEORGE ORWELL is the pen name of the British novelist and social critic who was born in India in 1903 as Eric Arthur Blair. He lived for most of his forty-seven years in Paris and London and is best known for his satirical novels *Animal Farm* (1946) and *Nineteen Eighty-Four* (1949). He also wrote many other books and essays in which he explored sociopolitical issues and questions of human freedom.

DONALD CULROSS PEATTIE was born in Chicago in 1898. After working as a government botanist in the early 1920s, he became a full-time freelance writer, producing fiction, nonfiction, and children's books. His writing fuses natural history with reflections on the human condition and the promise and perils of modern science. In addition to *An Almanac for Moderns* (1935), he is the author of *Flowering Earth* (1939), the autobiographical *The Road of a Naturalist* (1941), and numerous other works.

KATHRYN PHILLIPS is a journalist who has written for magazines such as *Omni, International Wildlife,* and *Discover.* In addition to *Tracking the Vanishing Frogs* (1994), she is the author of *Paradise by Design* (1998). She lives in Ventura, California.

PLINY THE ELDER aspired to completeness in compiling his thirty-seven-volume *Natural History,* including both firsthand observations and much secondhand information about astronomy, geography, anthropology, and especially zoology and botany. He lived in Italy during the first century C.E. and died of asphyxiation while investigating the eruption of Mt. Vesuvius in 79.

ROBERT MICHAEL PYLE migrated to the literary sphere from the world of conservation biology on the stylish wings of butterflies. After producing several indispensable guides to butterflies, he has focused his writing, which integrates ecology, environmentalism, and great affection, on dear but damaged lands, as in *Wintergreen* (1986), about his current home territory in southwestern Washington State, and *The Thunder Tree* (1993), about his childhood home in Colorado. His other books include *Where Bigfoot Walks* (1995) and *Chasing Monarchs* (1999).

JOHN RAY was the leading naturalist of the late seventeenth century. Inspired by the Baconian spirit of systematic observation, he contributed to the scientific revolution in biology through his extensive classifications of flora and fauna both in his native England and on the Continent, formulating a seminal concept of species that reflected his belief in nature's divine design.

DAVID SCOTT is a researcher at the Savannah River Ecology Laboratory in Aiken, South Carolina.

DALLAS LORE SHARP enjoyed a successful literary life that was neatly bisected by the turning of the twentieth century. Born in southern New Jersey in 1870, he graduated from Brown University and then from theological school at Boston University, where he returned to teach English from 1902 until his death, in 1929. He lived on a farm southeast of Boston and wrote many spirited works, including *A Watcher in the Woods* (1903), *The Face of the Fields* (1911), *Beyond the Pasture Bars* (1914), and *Sanctuary! Sanctuary!* (1926).

DAVID C. SMITH is a senior lecturer in the biology department at Williams College.

JAN SWAMMERDAM, along with his fellow seventeenth-century Dutchman Antoni van Leeuwenhoek, was a pioneering microscopist and made important studies of insects and many other animals, including influential investigations of amphibians. Although he was sympathetic to mystical religion, he maintained a preference for mechanistic explanations in his scientific work.

EDWIN WAY TEALE was one of the most well known and widely traveled American naturalists of the twentieth century. Over a span of fifty years, from 1930 until his death, in 1980, he produced some thirty books, many illustrated with his own photographs, and hundreds of articles, particularly for

Audubon magazine. His 100,000-mile ramble through the seasons and repeatedly across the American continent resulted in *North with the Spring* (1951), *Autumn Across America* (1956), *Journey into Summer* (1960), and *Wandering Through Winter* (1965), the last of which received a Pulitzer Prize.

J. ARTHUR THOMSON served as Regius Professor of Natural History at the University of Aberdeen in his native Scotland from 1899 until 1930, three years before his death. He produced scholarly works on evolutionary biology and philosophy, but he is especially remembered for his many popular books, including *What Is Man?* (1923) and *The Wonder of Life* (1927), and for his synthesis of science and religion.

HENRY DAVID THOREAU imbibed New England transcendentalism from Ralph Waldo Emerson and is best known for his experiment in simple living conducted from 1845 to 1847 on the shore of Walden Pond, at Concord, Massachusetts. His detailed journal of natural history and philosophy, which he began in 1837, formed the basis for most of his books and reveals the distinctive cadence of a life lived close to nature and at odds with the rising materialism of American science and society.

EDWARD TOPSELL, whose *History of Four-Footed Beasts* was the first major work on animals printed in the English language in Great Britain, was an English clergyman of the late sixteenth and early seventeenth centuries.

LAURIE J. VITT is curator of reptiles at the Sam Noble Oklahoma Museum of Natural History and a professor of zoology at the University of Oklahoma.

GILBERT WHITE is indelibly linked with the village of Selborne, in the south of England, where he was born in 1720 and died in 1793, having served as curate from the early 1750s. His *Natural History of Selborne* is one of the most frequently published books in the English language.

HENRY M. WILBUR is a professor of biology at the University of Virginia.

TERRY TEMPEST WILLIAMS was born in Nevada and has been shaped by her experiences in the Great Basin and the Colorado Plateau. Her numerous books, most notably *Refuge* (1992), *Desert Quartet* (1995), and *An Unspoken Hunger* (1995), speak of the greater awareness of life and death that grows out of an increased intimacy with nature. She has worked as naturalist-in-residence at the Utah Museum of Natural History but recently moved from Salt Lake City to a small town in southern Utah.

ANN HAYMOND ZWINGER combines the wonderfully complementary skills of exacting illustration and graceful writing. She began drawing as a child growing up in Indiana, though her love of nature came later, and her training in art history has given her a perceptive eye, which, now through both word and image, she tries to foster in her readers. Her many years of living in Colorado and her explorations throughout the West have provided the material for many of her books, including *Beyond the Aspen Grove* (1970), *Run, River, Run* (1975), *The Mysterious Lands* (1989), and *Shaped by Wind and Water* (2000).

Index

Abstractions, 133, 135
Academy of Sciences in Paris, 209
Acid rain, 186, 193
Adam's woods, 143–46
Adder snake, 120–21
Advertisement, 41–45
Aelian, 29
Aeromonas hydrophila, 188
Afoot in England (Hudson), 147
African clawed frog, 178–79
Age of machines, 130
Agnosticism, 67
Albert the Great, 12, 39, 207–208
   De Animalibus, 19–21
Albugo, cure for, 17
Aldrovandus, 27, 28
Alexander the Great, 54
Algonquin Park, Ontario, 190–92
Alienation of humanity from the natural world, 6–7, 10, 111, 134–35
   free enterprise and, 204–205
   Miller on, 204–13
   see also Reclaiming paradise; Remembering the earth
Allegheny Mountains, 145
Almanac for Moderns, An (Peattie), 124–27
Alytes obstericans, 20, 101, 106
Amatus Lusitanus, 29
Amblyrhynchus, 69
Ambystoma gracile, 148, 149–50, 180–81
Ambystoma opacum, 106–109
Amnion, 83
Amphibians, 1–10, 148
   as bioindicators, 175–76
   birth and evolution of, 168–69
   see also specific amphibians
"Amphibians as Harbingers of Decay," 161, 175–76
"Amphibians in a Bad Light," 176–81

Andes Mountains, 168, 185, 187
Animated nature, 54–55
"Animate Nature," 79
Anstis, M., 101
Antelopes, 54
Anthropocentrism, 8
Anura, 169
Aplin, Ken, 194–95, 197, 199–200
Apocynon, 25
Appalachian Mountains, 144
Aquinas, Thomas, 112, 207
Archetypes, 89
"Are Pathogens Felling Frogs?," 194–200
   death down under, 196–98
   revisiting U.S. dieoffs, 198–200
Aristophanes, 10
Aristotle, 2, 9, 11, 29, 35, 37
   deformities and, 206–207, 208
   Historia Animalium, 14–15
   on potentiality, 126–27
   on salamanders, 15, 25–26
Arizona, 196, 198
Arnold, Matthew, 113
Arrhenius, 127
Arroyo toad, 174, 184
Ascaphus truei, 151–53
Ashurbanipal, King, 205–206
Asteroid collision with the Earth, 169
Astronomical seasons, 132
Astronomy, 133, 135
Atelopus mucabajiensis, 185
Atelopus oxyxrhynchus, 185
Atelopus soriani, 185
Atherton Tablelands, Australia, 196
Atrazine, 193
Audubon, John James, 137, 141
Audubon Society, 170
"Audubon's Salamanders," 136–42
Australia:
   death down under, 196–200

238   Index

Australia (continued)
   gastric brooding frog, see Gastric brooding frogs
   motorbike frogs, 194–95
Automata, 67, 72–75

Babbitt, Bruce, 163
Bacon, Francis, 6, 39, 208, 210, 211
"Bahia Blanca," 69
Bakewell, Lucy, 137
Balance of nature, 77
Bald eagles, 138, 170
Bartram, John, 40–41, 46
Bartram, William, 41
   Travels Through North and South Carolina . . . , 45–49
Bataillon, 126
Bates, Henry Walter, 96, 103
Batesian mimicry, 96, 103
Batrachians, 69, 71, 85
Batrachochytrium dendrobatidis, 195–200
Battle between humanity and nature, 6–7
Beagle HMS, 5–6, 65, 68–70, 97–98
Beasely, Val, 199
Beauty, 125
Bedtime Story Books (Burgess), xiv
Bekoff, M., 100
Bell frog, 48
Berrill, Michael, 189, 192, 193
Bertolucci, Jaime, 185
Bible, the, 23–24, 95, 206
Bibron, M., 70
Biocentrism, 8–10
Biodiversity, 8–9
Biological pressure, 84
Biology:
   Huxley on, 81–88
   Thomson on, 76–81
Biology on the Seasons, The (Thomson), 67, 76–81
Birds, 170
   at courting-time, 86
Black bears, 47
Blacktail Canyon, 155–56
Black toad, 49
Blake, William, 147
Blaustein, Andrew R., 161, 199, 201
   "Amphibians in a Bad Light," 176–81

Boetius, 27
Book of a Naturalist, The (Hudson), 59–63
Book of Nature, The (Swammerdam), 31–33, 44
Borax toad, 20
Brain, molecular changes in the, 74–75
Brassavolus, 27
Brazil, 185
Breathing, 80–81, 148, 186, 191
"Breeding Congress, A," 105–109
British Columbia, 188, 194
Brongniart, Alexandre, 5
Brooding, 100–101
   gastric, *see* Gastric brooding frogs
   pouches for, 98, 101
Brooks, Ron, 190–91, 193
Brower, David, 147
Brower, Lincoln, 150
Browne, Sir Thomas, 2, 3, 4, 9, 209
   Pseudodoxia Epidemica, 13, 26–30
Brown pelicans, 170
Bruises, treatment for, 17
Buergeria buergeri, 187
Bufo, 19–20
   B. alvarius, xvi
   B. boreas, xiii–xv, 150–51, 176–81, 188, 198
   B. bufo, xvi
   B. marinus, xvi, 102
   B. microscaphus californicus, 174, 184
   B. periglenes, 168, 185, 187–88
Bull frogs, 151, 167
   Bartram on, 48
   Burroughs on, 55–56
   described, 190–93
   Leahy on, 189–94
   Thoreau on, 52
Bully for Brontosaurus: Reflections in Natural History (Gould), 95–105
Burgess, Thornton, xiv, 150
Burghardt, G. M., 100
Burroughs, John, 40, 41, 111
   Pepacton, 55–58
Bush, George, 173
Busse, K., 101
Butterfly mimicry, 96, 103, 150

Cady, Harrison, xiv
Caecilians, 169
Caldwell, Janalee P., 175–76

California, 164–65, 167, 194
California condors, 170
California red-legged frog, 172–75
Canada, 188, 189–94, 201
Cancer, cure for, 44–45
Cannibalism, 155, 156
Cardan, Jerome, 23
Carey, Cynthia, 188, 196
Carter, D. B., 99
Cascade Range, xiii, 176–81
Cascades frog, 168, 177–80
"Case of the Vanishing Frogs, The," 181–89
  gathering evidence, 183–86
  inconclusive evidence, 187–88
  interim recommendations, 188–89
  possible suspects, 186–87
  why we care about the victims, 182–83
Catbird, 125
Cattle, ingestion of frogs by, 17, 21
Caudata, 169
Causes, 206–208
Ceratophrys dorsata, 76
Chain of being, 5–6
"Charismatic megafauna, 170
Chemicals, industrial and agricultural, 162, 166, 186, 187, 192, 193, 201, 203, 204
Cherokee Indians, 144
Children's International Rainforest, 185
Chytrid fungus, 162, 195–200
Clamminess, 82
Climate change, 196
Clucking frog, 55, 56
Coldness, 82
Coleridge, Samuel Taylor, 41
Colombia, 187
Colorado, 188, 194, 196, 198, 215–16
Colorado River, 155–56, 158
Colorado River toad, xvi
Committee on the Status of Endangered Wildlife in Canada (COSEWIC), 193, 194
Common frog, 22, 42
Common mergansers, 138
Common toad, xvi, 62, 128–30
  Orwell on, 128–30
Competitiveness, 65
Compromises, 82

Consciousness, 73–75
Conservation Biology, 185
"Contributions to an Insect Fauna of the Amazon Valley," 96–97
Corben, C. J., 98, 101
Corn borer, 123
"Cosmic environmentalism," 12–13
Cosmos, 87–88
  see also Interpreting the cosmos
Costa Rica, 168, 195, 197
  Monteverde, 184–85, 187–88
Cowbirds, 121
Craftsmanship of nature, 206–207
Creationism, 96–97
Croaking:
  Albert the Great on, 20
  Aristotle on, 14
  Burroughs on, 55–56
  Hudson on, 61–62
  Huxley on, 82, 85–86, 87
  Leahy on, 189, 191–92
  Pliny the Elder on, 16–17
  Pyle on, 151
  Thomson on, 76
  Thoreau on, 52–55
  Topsell on, 24
  Zwinger on, 154, 155–56, 215–16
Crows, 121–22
Crump, Martha, 185
Cryptobranchus alleganiensis, 145
Cryptogamia, 43
Cutaneous absorption, 69
Cutaneous respiration, 80–81, 82, 148, 186, 191
Cutworms, 119, 123
Cuvier, Georges, 5

"Dance of the Frogs, The," 68, 88–95
Darwin, Charles, 5–6, 41, 65, 125, 210
  mimicry and, 96–97, 103–104
  Origin of Species, 6, 65, 67, 70–71, 96
  Voyage of the Beagle, 68–70
Darwinism, 67, 68
Daszak, Peter, 196, 197
David and Goliath, 95
"Day of the Peepers, The," 131–36
DDT, 187
De Animalibus (Albert the Great), 19–21
Declines and deformities:
  early indications of, 161–63

Declines and deformities (*continued*)
  Minnesota school children and, 162, 193, 201, 203, 211–12
  see also Reading the Signs of the Times
Declining Amphibian Populations Task Force (DAPTF), 161, 163, 182
Defoe, Daniel, 217
Democritus, 17
Dendrobates pumillo, 183
Derham, William, 36
Descartes, René, 39, 72, 74, 209
Desert Quartet (Williams), 157
De Subtilitate (Cardan), 23
Detergents, 186
Developmental processes, 76–81
Dillard, Annie, 112
  Pilgrim at Tinker Creek, 142–46
"Dimensions of Deformity," 203–13
Dinosaurs, 169
Dioscorides, 29
Discontentment, 130
Diseases, 188
  cures for, 17, 26, 44–45
  fungal, 195–200
  theory of, 79
Disjuncture, 152
DNA:
  fungal, 197
  pesticides and, 193
  UV-B and, 178–80, 187
Douthat State Park, 145
Downcanyon: A Naturalist Explores the Colorado River Through the Grand Canyon (Zwinger), 155–56
Dreyer, Albert, 88–95
Driesch, Han Adolf Eduard, 126
Drost, Charles, 184
Dualism, *see* Alienation of humanity from the natural world
Du Bois, 70
Dunn's salamander, 148, 179

Earl of Bedford, 25
Early naturalists, *see* Interpreting the cosmos
Early warning systems, 163, 175–76, 189, 193, 194, 195
Easter, 131–34
Ecological science, 111–12, 210–13

Ecological sensibility, 212–13
Ecology, 65, 210
Ectothermism, 191
Ecuador, 185, 187
"Edge of Night, The," 113–18
Egg laying, *see* Reproduction
Egypt, ancient, 3, 22, 205–206, 207
Ehrlich, Paul, 201
Einstein, Albert, 143
Eiseley, Loren, 68
Eleutherodactylus coqui, 127
E. jasperi, 185
Ellis, John, 44
Endangered Species Act, nomination process for, 172–75
Endocrine system, 193
Energy, conveyors of, 9
Environmental Defense Center, 174
Environmental impact assessments, 188
Environmentalism, 210
Erasmus, 35
Essays in Popular Science (Huxley), 81–88
Estrogens, environmental, 187
Ethics, 208, 210
European common frog, 22, 42
Eurycea longicauda longicauda, 139
Evolution, 65–68, 111
  brooding and, 100–102
  fortuitous change and, 103–104
  Huxley on, 84
  mimicry and, 96–97, 103–104
Explorers Club, 88
Extinction, 84, 185, 194, 198, 217
  disjuncture and, 152
  of gastric-brooding frog, 100, 104–105, 167, 183, 197
  mass, 169, 195
  see also "Reading the Signs of the Times"
Eyes, 128

Face of the Fields, The (Sharp), 113–18
Fairies, 122–23
Federal Register, 174
Fellers, Gary, 184
Fiddler-crab, 87
First World Congress of Herpetology, 161, 182, 184
Fluctuating water levels, 194
Foothill yellow-legged frog, 165–67

Formation of Vegetable Mould Through the Action of Earthworms, The (Darwin), 65
Fortuitous change, 103–104
Fragmentation, 210–13
Free enterprise, 204–205
"Frog and Biology, The," 81–88
Frogs:
    described, 20–24
    first, 169
    as lacking advocates, 170
    toads compared to, 169
    see also specific frogs
"Frogs, Toads, and the Toadstone, Of," 26–30
Frog skin, 183
Frog time, 91–95
Frogwatch USA, 163
Froman, Robert, 215
Fromondus, 34, 35, 36, 37
Fungi, 162, 195–200

Gaia theory, 9
"Galápagos Archipelago, 69–70
Galen, 29, 30
Galileo, 12, 209
Game fish, 187, 193
Gametes, 96
Gardiner, David, 202
Gastric brooding frogs, 68
    extinction of, 100, 104–105, 167, 183, 197
    Gould on, 95–105
    Halliday and Heyer on, 183, 184
    Phillips on, 167
Gastric ulcers, 183
General History of Plants (Ray), 4
Gesner, Conrad, 12, 27, 54
Giant water bug, 142–43
Gilbert and Sullivan, 104
Ginger's Bay, South Carolina, 105–109
Glacken, Clarence, 12
Goblins, 123
God, 46, 143, 183
Golden toad, 168, 185, 187–88
Gould, Stephen Jay, 68
    Bully for Brontosaurus: . . . , 95–105
Grahame, Kenneth, xiv, 150
Grand Canyon, 155–56, 158, 216
Gray's River, Washington, 146–52

Great Basin, 154, 194
Great Lakes region, 198
Great Salt Desert, 153–55
Grebes, 86
Green, David, 181–82, 193, 194
Green-eyed tree frog, 197
Green frog, 21, 28, 48
Gregory, Pope, 133
Grinnell, Joseph, 184
Grosbeaks, 125, 138
Guns, 47, 122
Gymnophiona, 169
Gynilus, 28–29
Gyrini (gyrinos), 23

Habitat destruction, 161–62, 170, 199
    Halliday and Heyer on, 184–85
Haeckel, Ernst, 65–67, 210
Halliday, Timothy R., 161
    "Case of the Vanishing Frogs," 181–89
Harvey, William, 76
"Has a Frog a Soul?," 67
Hayes, Marc, 171–74
Hayes, Tyrone, 187
Hays, John, 178–79
"Heaven and Earth in Jest," 142–43
Hellbender, 145
Herbicides, 186, 192, 193
"Here Goes Nothing," 95–105
Hermaphroditism, 149
Herons, 86
Herrick, Robert, 82
Heyer, W. Ronald, 161
    "Case of the Vanishing Frogs," 181–89
Hibernation, 21, 57–58, 154
High land frogs, 49
History of Animals (Gesner), 12
History of deformities, 205–13
History of Four-Footed Beasts and Serpents, The (Topsell), 12, 21, 22–26, 54
History of Insects, The (Swammerdam), 31–33
Hoffman, Peter, 179
Homing instinct, 115–16
Horned toads, 20
Horse-nails, 23
Hovingh, Peter, 154
Howes, G. B., 98
Hudson, William Henry, 40, 41, 147
    Book of a Naturalist, 59–63

Human-created environmental disasters, 169–70
  types of, 172
Human transformation of nature, *see* Reclaiming paradise
Huxley, Julian, 67
  Essays in Popular Science, 81–88
Huxley, Thomas Henry, 6, 65
  "Hypothesis That Animals Are Automata," 67, 72–75
Hyla, 58, 69, 114
H. crucifer, see Spring Peepers
H. femoralis, 187
H. pacifica, xv
H. regilla, 151, 177–80, 198, 200
Hylodes, 52–53

"Ideagenous molecules," 74–75
Imagination, 54–55
Immense Journey, The (Eiseley), 68
Imperturbability, 51
Incipient stages, problem of, 97, 103–104
Indicator species, 163, 175–76, 183, 193, 194, 195
Ingram, Glen J., 98, 99, 101
"In here" and "out there," 204–13
Insects, 87
Interpreting the cosmos, 11–38
  Albert the Great, 19–21
  anonymous, second century C.E., 18–19
  Aristotle, 14–15
  Browne, 26–30
  introduction, 11–12
  Pliny the Elder, 16–17
  Ray, 33–38
  Swammerdam, 31–33
  Topsell, 22–26
"Intuitive perception," 210
Irvine, California conference of 1990, 161, 171, 175–76, 182
Irvine, William, 67
Isolation of populations, 152, 156

Jacob, François, 13
James, Timothy, 197
Japanese tree frogs, 187
Javan flying frog, 136
Jennings, Mark, 164–66, 169, 171–75

Jennings, Ray and Helen, 164–66
Johnson, Pieter, 201–203
Journal (Thoreau), 50–55
Juan Fernández Islands, 216–17
Judeo-Christianity, 183
Judge, Kevin, 191, 193

Kaiser, Jocelyn, 162
  "Trematode Parasite Causes Some Frog Deformities," 200–203
Keratin, 198
Kern River, California, 164–65, 167
Kingsley, Charles, 140
Kiranides, 26
Klickitat Hatchery, 152
"Knowledge" of nature, 208, 210
Koch's postulate, 197
Koran, the, 143
Krutch, Joseph Wood, 111, 112
  Twelve Seasons, 131–36

Lacerta, 44
Lake Sasajewun, Ontario, 190–91
Lake Scugog, Ontario, 189
LaMarca, Enrique, 185
Lamberton, Ken, xvi
Land-eft, 44
"Land-frogs," 18–19, 22, 49
Language, 14
Lannoo, Michael, 186, 199, 201, 203
Latreille, Pierre-André, 5
"Lazarus," 164–71
Leahy, Stephen, 162
  "Sound of Silence," 189–94
Leisure, 130
Lembus, 23
Leopard frogs, 167, 168, 194, 196, 198, 215–16, 217
Leopold, Aldo, 210, 212
Lepidosiren, 77
Life, 127, 130, 134
Linnaeus, Carolus, 4–5, 40, 41, 44
Lipps, Karen, 197
Lizards, 69, 70
Lobefin fish, 168
Locust trees, 144
Loeb, Jacques, 126
Longcore, Joyce, 197, 199–200
Long-tailed salamanders, 138–41

Long-toed salamanders, 148
Lost Lake, Oregon, 176–77
Lovejoy, Arthur O., 3
Love of Nature, 130
Lowcock, Les, 193
Lowell, James Russell, 40
Lowland leopard frogs, 196, 198
Lunde, Kevin, 202
Lungfish, 168
Lupus marinus, 28

MacCaig, Norman, 215
McDiarmid, R. W., 100
McDonald, K. R., 99
McPhee, John, 149
Magical uses of frogs and toads, 12, 13
   Albert the Great on, 21
   Browne on, 29
   Pliny the Elder on, 17
   Topsell on, 24
   White on, 44–45
"Magic eye," 211
Mammals, 170
Man on the Beasts (Albert the Great), 19–21
Marbled salamander, 68
   life cycles of, 105–109
Marcellus, 22
Marine toad, xvi
Más a Tierra, 216–17
Massachusetts, 216
Massarius, 22
Mathiolus, 29
Mating, *see* Reproduction
Matter, 87–88
Mauritius, 70
Maxwell, Ed, 151, 152–53
Mechanistic biologists, 126–27
Medicalization of anomalies, 209
Medicines, 183
   pseudo-, 17, 21, 24, 26, 44–45
Medieval Europe, 146, 207–208
Merret, 44
Mesopotamia, 206, 207, 211
Metabolism, 191
Metamorphosis, 77–85, 192
Meteyer, Carol, 203
Methoprene, 186
Midwest, 201, 203
Midwife toad, 20, 101, 106

Migration:
   of frogs, 43–44
   of salamanders, 105–107
   of toads, 59–63
Miller, Gordon L., "Dimensions of Deformity," 203–13
Mill Grove, Pennsylvania, 137–39
Mimicry, 96–97, 150
   initiation of, 97, 103–104
Mine Run, Pennsylvania, 138–39
Minnesota, 162, 193, 201, 203, 211–12
Missing links, 5
Modern man, 134–35
Mohr, Charles E., 139
Monarch butterfly, 96, 150
Montgomery County Park System, 137
Moon, the, 144
Morality, 46–47, 171, 182–83, 208
More, Henry, 3, 9
Morell, Virginia, 162
   "Are Pathogens Felling Frogs?," 194–200
Motorbike frogs, 194–95
Mountain yellow-legged frogs, 198
Mount Rainier National Park, 153
Mud iguana, 44
Murder, 47
Museum of Natural History, Paris, 5, 209
Mysterious Lands: A Naturalist Explores the Four Great Deserts of the Southwest, The (Zwinger), 153–55
"My Twenty-Four-Dollar Toad," 118–23

Naselle Salmon Hatchery, 151, 153
National Academy of Sciences, Board of Biology, 182
National Research Council, 104–105
   Board on Biology, 175
National Science Foundation, 176
National Wildlife Federation, 170
Natterjack, 62
Natural History (Pliny the Elder), 16–17
Natural History of Selborne, The (White), 40, 41–45, 65
Natural History Reviews, 96–97
Naturalist on the River Amazon, The (Bates), 96
Natural selection, 65

Nature:
    historical views of, 204–209
    see also Alienation of humanity from the natural world
Near East, ancient, 205–206
New Mexico, 198
Newts, 44, 45
    Dillard on, 144–46
New York Times, The, 171
Nicander, 29
Nichols, Donald, 196, 197, 198
Nid, John, 34
Nisqually River, Washington, 153
Non-native species, 162, 187, 194
North American Amphibian Monitoring Program (NAAMP), 163
North American Reporting Center for Amphibian Malformations (NARCAM), 163
Northern cricket frog, 199
Northern leopard frog, 167, 194, 196, 198
Northern toads, xiii–xv, 150–51, 198
Northwestern salamanders, 148, 149–50, 180–81
Northwood, J. d'Arcy, 137–38, 140, 141
Nutrition, 80

Objectification, 208–13
O'Brien, P., 102–103
Oceanic islands, 69–70, 71
Oecologie, 65
Olympic salamanders, 148
Omens, 205–208, 211
Ontario, 189–94
Ontario Ministry of Natural Resources (OMNR), 190
"On the Hypothesis that Animals are Automata, and Its History," 67, 72–75
Ophthalmia, cure for, 17
Origin of Species, The (Darwin), 6, 65, 67, 70–71, 96
Orwell, George, xvi, 112
    "Thoughts on the Common Toad," 128–30
Osteichthyes, 168
Otters, 191
Owls, 114

Ozone depletion, 161
    UV-B radiation and, 178–81, 186–87, 201

Pacific tree frogs, 151, 177–80, 216
    parasitic infection of, 198, 201, 202
Panama, 195, 197
Parasites:
    chytrid fungus, 195–200
    trematode flatworm, 162, 200–204
Parental and filial affection, 47
Parthenogenetical development, 126–27
"Particular Treatise on the Generation of Frogs, A," 31–33
Pascal, Blaise, 143
Patience (Gilbert and Sullivan), 104
Patuxent Wildlife Research Center, 163
Paulson, Dr. Dennis, 148
Payne, John Howard, 115
Peattie, Donald Culross, 112
    Almanac for Moderns, 124–27
Pennant, Thomas, 43–45
Pepacton (Burroughs), 55–58
Perault, Monsieur, 35
Peregrine falcons, 170
Perkiomen Creek, Pennsylvania, 137–39
Pesticides, 162, 166, 186, 187, 192, 193, 201, 203, 204
Petranka, Jim, 108
Phenomena, 133
Philarchus, 23
Phillips, Kathryn:
    "Lazarus," 164–71
    "Place for Frogs," 171–75
    Tracking the Vanishing Frogs, 163, 164–75
Phoebes, 138
Photolyase, 179, 180
Phryniscus nigricans, 69
Physiologus, 12, 18–19
Pierius, 29
Pignorius, 28
Pilgrim at Tinker Creek (Dillard), 142–46
Pine woods tree frogs, 187
Pioneering nature writers, *see* Reclaiming paradise
"Place for Frogs, A," 171–75
Plagues of Egypt, 23–24, 206
Planorbid snails, 201, 204
Plato, 2, 9

Plethodon dunni, 148, 179
Pliny the Elder, 4, 11, 12, 23, 28, 29, 37, 38
  Natural History, 16–17
Poison dart frogs, 197
Poison-frog toxins, 183
Poisons, antidote for, 17, 21, 24
Pollutants, 162, 186, 204
Polypterus, 77
Pond frogs, 124
Pope, 75
Porta, 27
Porter, David, 197
Porwigle, 28–29
"Postmetamorphic death syndrome," 198–99
Potentiality, 126–27
Pounds, Alan, 185
Power of Toads, The (Rogers), xiii
Praying mantises, 144–45
Primitive religions, 88–90
Proceedings of the Zoological Society of London, 98
Process of life, pleasure in, 130
Proctor, Noble, 153
Propagation, *see* Reproduction
Prostaglandin, 103–104, 183
Protoamphibians, 104
Protopterus, 77
Psammophis Temminckii, 70
Pseudodoxia Epidemica (Browne), 13, 26–30
Puerto Rican *coqui,* 127
Puerto Rico, 168, 185, 200
Pyle, Howard Whetstone, xv
Pyle, Robert Michael, xiii–xiv, 112
  Wintergreen: Listening to the Land's Heart, 146–53
Pyle, Thea, 153

Quebec, 193
Queensland, Australia, 194–95, 196–97
"Quiet desperation," 40
Qulu, 154

Rain, 147
Rainbow Bay, South Carolina, 108
Raining frogs, 23
  Ray on, 33–37, 44
Rana, 20

R. arborea, 28, 44
R. aurora, 151, 177
R. boylii, 165–67
R. cascadae, 177–80
R. catesbeiana, see Bull frogs
R. fontinalis, 51
R. halecina, 51, 52, 53
R. marina, 22
R. mascariensis, 70
R. muscosa, 182
R. palustris, 52, 53
R. pipiens, 167, 194, 196, 198
R. temporaria, 42, 76–78
Ranids, 166–67
Ranunculus viridis, 28
Ray, John, 4, 13, 209
  Wisdom of God . . . , 33–38, 39–40, 43–44
"Reading the Signs of the Times," 161–213
  Blaustein, 176–81
  Halliday and Heyer, 181–89
  introduction, 161–64
  Kaiser, 200–203
  Leahy, 189–94
  Miller, 203–13
  Morell, 194–200
  Phillips, 164–75
  Vitt et al., 175–76
Reagan, Ronald, 173
Reclaiming paradise, 39–63
  Bartram, 45–49
  Burroughs, 55–58
  Hudson, 59–63
  introduction, 39–41
  Thoreau, 50–55
  White, 41–45
Red-backed salamanders, 148
Red-eyed vireo, 121
"Red leg" disease, 188
Red-legged frogs, 151, 177
Red-spotted toad, xvi, 155–56
Red toad, 49
Relicts, 152
Religio Medici (Browne), 2
"Remembering the Earth," 111–59
  Dillard, 142–46
  introduction, 111–12
  Krutch, 131–36
  Orwell, 128–30

"Remembering the Earth," (*continued*)
  Peattie, 124–27
  Pyle, 146–53
  Sharp, 113–23
  Teale, 136–42
  Williams, 157–59
  Zwinger, 153–56
Rennaissance, the, 207–208
Reproduction:
  Albert the Great on, 20–21
  Aristotle on, 14–15
  Bartram on, 49
  Blaustein on, 177–81
  Browne on, 28–29
  Gould on, 95–105
  Halliday and Heyer on, 183
  Huxley on, 82–87
  Leahy on, 192, 194
  Orwell on, 128–30
  Peattie on, 124–27
  Pliny the Elder on, 16
  Pyle on, 148–53
  Ray on, 33–38
  Scott on, 105–109
  Swammerdam on, 31–33
  Teale on, 140–41
  Thomson on, 76–81
  Topsell on, 22–23
  White on, 43–44
  Zwinger on, 154–55
Reptiles, 43
  advantage gained by, 84
  on oceanic islands, 69–70
Respiration, 80–81
  cutaneous, 80–81, 82, 148, 186, 191
Responsible citizenship, 210–13
Retinoic acid, 186
Retinoids, 202
"Revealing secrets," 17, 21, 24
Rhacophorus reinwardti, 136
Rheobatrachus silus, *see* Gastric brooding frogs
R. vitellinus, 100
Rhinoderma darwini, 98, 102
R. rufum, 101
Ribeiroia, 201–203
Richardson, Frank, 152
"Rio de Janeiro," 68–69
Ritchie, Euan, 202
Roadways, 193

Robinson Crusoe Island, 216–17
Rogers, Pattiann, xiii
Romans, ancient, 146
Rough-skinned newts, 148, 150
Rowley, Anthony, 86
Royal Society of London, 5, 6, 209
"Rubeta" ("rubetum") frog, 21

St. Vincent, Bory, 70, 71
Salamanders, 148–50, 184
  fire-related myths, 3–4, 15, 16, 18, 25–26, 29–30, 146
  poison myths, 146
  UV-B and, 179–81
  see also specific salamanders
Salamandra aquatica, 44
Sanctuary! Sanctuary! (Sharp), 118–23
Santa Paula Creek, California, 167
Sarenus Sammonicus, 29
Savanna cricket, 48–49
Savannah River Ecology Laboratory, 108
Scaliger, 30
Scaphiopus, 154–55
Schaller, George, 171
Science, 40–41
Science, 176, 195, 200n, 201
Science News, 105
Scientific American, 150
Scientific essayists, 65–109
  Darwin, 68–71
  Eiseley, 88–95
  Gould, 95–105
  J. Huxley, 81–88
  T. Huxley, 72–75
  introduction, 65–68
  Scott, 105–109
  Thomson, 76–81
Scientific method, 208–13
Scott, David, 68
  "Breeding Congress," 105–109
Scott, Norman, 198–99
Seasons, biology of, 67, 76–81
Sea-turtles, 70
Secretary of the Interior, 173
Selkirk, Alexander, 216–17
Sessions, Stanley, 202
Sex, 124–25
  see also Reproduction
Sexiness, 128
Sex Life of the Animals, The (Wendt), 149

Sextius, 29
Shad frog, 49
"Shaking tent rite," 89
Sharp, Dallas Lore, 112
   Face of the Fields, 113–18
   Santuary! Sanctuary!, 118–23
Shearman, D., 102–103
Sierra Club, 147
Sierra Nevada, 181–82, 198
"Skipping," 91–95
Slugs, 147–48
Smith, David C., 175–76
Smithsonian Institution, Biodiversity Programs Office, 161
Snails, 201, 204
Snakes, 16, 70, 82, 120–21
   tame, 63
Songbirds, 170
Sonoran Desert toad, xvi
Soul, 9–10, 67, 75
"Sound of Silence, The," 189–94
Southern cricket frog, 199
Southern toad, 107
Spadefoot toads, xvi, 154–55, 194
"Spadefoot Toads and Twilight, Of," 153–55
Spawning, *see* Reproduction
Speare, Richard, 196, 198, 199–200
Speckled frog, 48
Spiders, 20, 30, 87
   resistance to the poison of, 25
Spiders, Snakes, and Other Outcasts (Froman), 215
Spinal cord, 72–73
Spingelius, 28
Spontaneous generation, 15
   Ray on, 33–38
Spotted owls, 170
"Spring," 143–46
Spring equinox, 132
Spring peepers, 52, 53
   Burroughs on, 55, 56, 58
   Dreyer on, 92
   Krutch on, 131–36
   Peattie on, 125
   Zwinger, 215, 216
Springtime, 128–30
   early, 124–27
   Eiseley on, 90
   late, 92

Sredl, Michael, 198
Stages of development, *see* Reproduction
Star Thrower, The (Eiseley), 88–95
Stoat, 60
Storer, Tracy, 184
Strawberry poison frog, 183
Striped beetle, 123
Survival of older life forms, 84
Swammerdam, Jan, 4, 13, 43
index:Swanson, Sarah, 191
Swede Park, Washington, 146–47
Sweet, Sam, 174
Swifts, 138
Symbolism, 206–208, 209, 211
Systema Naturae (Linnaeus), 44

Tadpoles:
   Albert the Great on, 20–21
   Browne on, 28–29
   Gould on, 101
   Huxley on, 83
   Leahy on, 192
   Orwell on, 128–29
   Pyle on, 152–53
   Ray on, 34, 36
   Thomson on, 76–81
   Topsell on, 22–23
   White on, 44
   Zwinger on, 155, 156
Tailed frog, 151–53
"Tale of Tadpoles," 67, 76–81
Tame creatures:
   frogs, 50–51
   snakes, 63
   toads, 62–63
Taskforce on Amphibian Declines and Deformities (TADD), 163
Teale, Edwin Way, 112
   Wandering Through Winter, 136–42
Teleological philosophy of nature, 208
Telling naturalistic tales, *see* Scientific essayists
Temporarioe, 28
Tendency to vary, 103–104
Thermometers, 53, 54
Thinking, 117–18
Thompson, Shaun, 190
Thomson, J. Arthur, *Biology of the Seasons,* 67, 76–81

Thoreau, Henry David, 40, 41, 68, 111, 137
  Journal, 50–55
"Thoughts on the Common Toad," 128–30
Thunder Tree, The (Pyle), xiv
"Toad as Traveller," 59–63
Toads:
  described, 19–20, 24–25
  frogs compared to, 169
  as lacking advocates, 170
  see also specific toads
Toad-stone, 20, 26–28
Toes, 58, 69
Tolerance, 147
Tongue, 14, 16–17, 49, 80
  of Bull frogs, 190–91
Topsell, Edward, 11, 41
  History of Four-Footed Beasts and Serpents, 12, 21, 22–26, 54
Tortoises, 70
Tracking the Vanishing Frogs, 163, 164–65
Transactions of the Linnaean Society, 96
Transcendental ecology, 11–13
"Travels of Georgia," 149
Travels Through North and South Carolina, Georgia, East and West Florida (Bartram), 45–49
Tree frogs, 21, 28, 44, 86
  green-eyed, 197
  Japanese, 187
  Pacific, *see* Pacific tree frogs
  pine woods, 187
  tropical, 216–17
  vocalizing of, 155–56
Tree of life, 6
Tree-toad, 55–58
  Sharp on, 113–18
"Trematode Parasite Causes Some Frog Deformities, A," 200–203
Trematode parasites, 162, 200–204
Trolls, 123
Tropical Science Center of Costa Rica, 185
Tropical tree frogs, 216–17
Trout, 187, 193
"True frogs," 166–67
Tully, 24
Turtledove, 60
Twelve Seaons, The (Krutch), 131–36

Twentieth-century nature writers, *see* Remembering the earth
Twilight, 118
Tyler, Michael J., 98, 99, 102, 104, 105, 186

Ultraviolet radiation (UV-B), 196, 201, 204
  Blaustein on, 161, 178–81
  Halliday and Heyer on, 186–88
United Kingdom, 184
U.S. Department of the Interior, 173
U.S. Fish and Wildlife Service, 172–74, 182
U.S. Forest Service, 171
U.S. Geological Survey, 163
U.S. National Park Service, 182

Value, 118–23
Venezuela, 185, 187
Venom:
  antidote against, 21
  benefits of, 183
  color and, 150
  northern toad, 150
  northwestern salamander, 146, 150
  rough-skinned newts, 150
  toads, 24, 43
Vermont, 201
Viceroy butterfly, 96, 150
Vision improvement, 17
Vitalistic view of life, 127
Vitt, Laurie J., 175–76
Viviparous frogs, 185
Voice, 14
Volcan Tajumulco, Guatemala, 184
Volition, 75
Voyage of the Beagle, The (Darwin):
  "Bahia Blanca," 69
  "Galápagos Archipelago," 69–70
  "Rio de Janeiro," 68–69

Wahkiakum County, Washington, 146–53
Wake, David, 161, 181–82, 196, 199, 203
Walden Pond, 50–55, 137
Wallace, Alfred Russel, 96
Wandering Through Winter (Teale), 136–42
"Water," 157–59
Water Babies, The (Kingsley), 140

Water-eft, 44, 45
"Water-frogs," 18–19, 22, 24, 28–29, 49
"Waterproof Wildlife," 147–53
Water-vole, 60, 61
Weather instruments, reptiles as, 53, 54
Web of life, 9–10, 12
Wellman, Bill, 144
Wells, H. G., 85
Wendt, Herbert, 149
Western pond turtle, 174
Western toads, 168, 176–81, 188
Whippoorwills, 116
White, Gilbert, 7, 67
   Natural History of Selborne, 40, 41–45, 65
Wilbur, Henry M., 175–76
Wilderness and Razor Wire (Lamberton), xvi
Willapa Hills, Washington, 146–53
Williams, Terry Tempest, 112, 154
   Desert Quartet, 157–59
Wilson, Edward O., 68
Wind in the Willows, The, xiv
Wintergreen: Listening to the Land's Heart (Pyle), 146–53

Wisconsin, 193
Wisdom, 117
Wisdom of God Manifested in the Works of the Creation, The (Ray), 33–38, 39–40, 43–44
Wood, Elijah, 51–52
Wood frogs, 50–51, 53–54, 55, 124, 215
Wordsworth, William, 41
World Conservation Union, Species Survival Commission, 161, 182
World Heritage rainforest parks, 196
World Wide Fund for Nature, 185
Worrest, Robert, 178
Wyoming, 198

Xenopus, 178–79

Yosemite National Park, 184
Yosemite toads, 168

Zwinger, Ann Haymond, 112
   Downcanyon: . . . , 155–56
   Epilogue, 215–18
   Mysterious Lands: . . . , 153–55

Island Press Board of Directors

SUSAN E. SECHLER, *Chair*
Vice President, Aspen Institute

HENRY REATH, *Vice-Chair*
President, Collector's Reprints, Inc.

DRUMMOND PIKE, *Secretary*
President, The Tides Foundation

ROBERT E. BAENSCH, *Treasurer*
Professor of Publishing, New York University

CATHERINE M. CONOVER

GENE E. LIKENS
Director, The Institute of Ecosystem Studies

DANE NICHOLS
Chairman, The Natural Step, U.S.

CHARLES C. SAVITT
President, Center for Resource Economics/Island Press

VICTOR M. SHER
Environmental Lawyer

PETER R. STEIN
Managing Partner, Lyme Timber Company

RICHARD TRUDELL
Executive Director, American Indian Resources Institute

WREN WIRTH
President, The Winslow Foundation